I0494094

Disclaimer

The publisher of this book is by no way associated with the National Institute of Standards and Technology (NIST). The NIST did not publish this book. It was published by 50 page publications under the public domain license.

50 Page Publications.

Book Title: Database-Assisted Design for Wind: Concepts, Software, and Example for High-Rise Reinforced Concrete Structures

Book Author: Dong H. Yeo;

Book Abstract: Time-domain analyses of wind effects on high-rise structures have been made possible in recent years by advances in wind pressure measurement and computer technology. Time domain solutions not only provide full phase information on structural responses to wind but can also account naturally for modes of vibration of any shape, including any number of higher modes of vibration, as well as for mode coupling. This study applies the Database-Assisted Design (DAD) methodology to the design of reinforced concrete high-rise structures. Given (a) the time histories of pressures, measured in the wind tunnel at a sufficient number of taps on the exterior faces of the building envelope for a sufficient number of mean speed directions, and (b) a preliminary design of the building, it is possible to calculate response databases for the demand-to-capacity indexes, inter-story drift, and top floor accelerations, that is, databases of responses induced by wind with any specified speed and direction. These responses are functions of the building s aerodynamic, geometric, structural, and dynamical features and are independent of the wind climate. The response databases are used in conjunction with a wind climatological database typically obtained by Monte Carlo simulation from measured extreme wind climatological data. The design is performed iteratively until the peak responses satisfy the design specifications. The present study is the first to apply Database-Assisted Design techniques to reinforced concrete high-rise buildings, and is in our opinion superior to conventional approaches currently in use from the point of view of physical modeling, accuracy, transparency, and convenience to the designer.

Citation: NIST TN - 1665

Keywords: Database-Assisted Design (DAD), mean recurrence interval, reinforced concrete, time-domain analysis, wind effects.

National Institute of Standards and Technology
Technology Administration, U.S. Department of Commerce

NIST TECHNICAL NOTE 1665

Database-Assisted Design of High-Rise Reinforced Concrete Structures for Wind:

Concepts, Software, and Application

DongHun Yeo

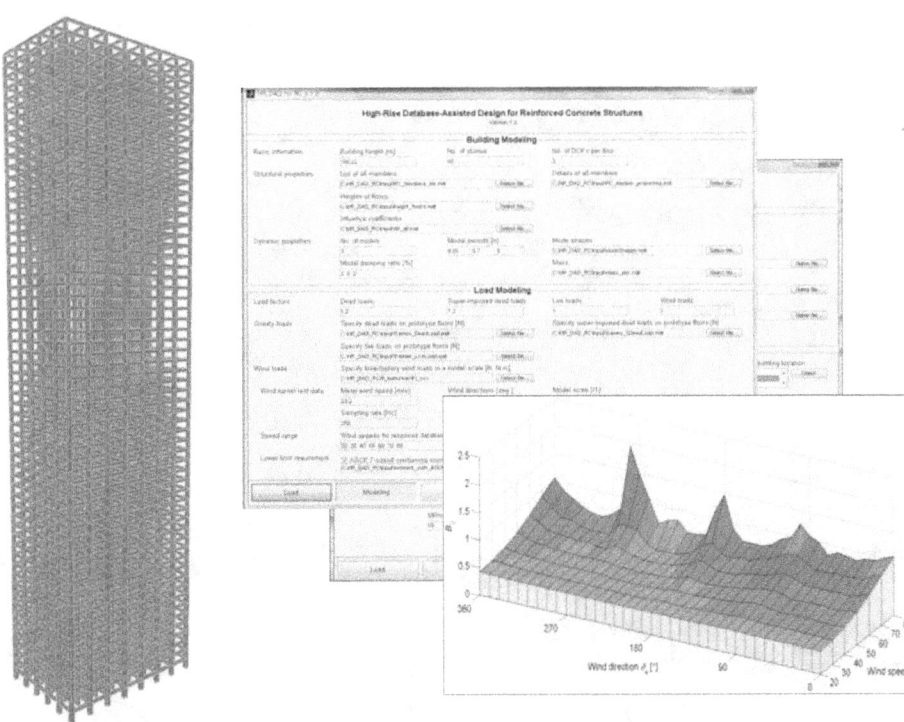

NIST TECHNICAL NOTE 1665

Database-Assisted Design for Wind:

Concepts, Software, and Example for High-Rise Reinforced Concrete Structures

DongHun Yeo

Building and Fire Research Laboratory
National Institute of Standards and Technology
Gaithersburg, MD 20899-8611

May 2010

U.S. Department of Commerce
Dr. Gary Locke, *Secretary*

National Institute of Standards and Technology
Dr. Patrick D. Gallagher, *Director*

Disclaimers

(1) The policy of the NIST is to use the International System of Units in its technical communications. In this document however, works of authors outside NIST are cited which describe measurements in certain non-SI units. Thus, it is more practical to include the non-SI unit measurements from these references.

(2) Certain trade names or company products or procedures may be mentioned in the text to specify adequately the experimental procedure or equipment used. In no case does such identification imply recommendation or endorsement by the National Institute of Standards and Technology, nor does it imply that the products or procedures are the best available for the purpose.

Abstract

Time-domain analyses of wind effects on high-rise structures have been made possible in recent years by advances in wind pressure measurement and computer technology. Time domain solutions not only provide full phase information on structural responses to wind but can also account naturally for modes of vibration of any shape, including any number of higher modes of vibration, as well as for mode coupling.

This study applies the Database-Assisted Design (DAD) methodology to the design of reinforced concrete high-rise structures. Given (a) the time histories of pressures, measured in the wind tunnel at a sufficient number of taps on the exterior faces of the building envelope for a sufficient number of mean speed directions, and (b) a preliminary design of the building, it is possible to calculate response databases for the demand-to-capacity indexes, inter-story drift, and top floor accelerations, that is, databases of responses induced by wind with any specified speed and direction. These responses are functions of the building's aerodynamic, geometric, structural, and dynamical features and are independent of the wind climate. The response databases are used in conjunction with a wind climatological database typically obtained by Monte Carlo simulation from measured extreme wind climatological data. The design is performed iteratively until the peak responses satisfy the design specifications.

The present study is the first to apply Database-Assisted Design techniques to reinforced concrete high-rise buildings, and is in our opinion superior to conventional approaches currently in use from the point of view of physical modeling, accuracy, transparency, and convenience to the designer.

Keywords: Database-Assisted Design (DAD), mean recurrence interval, reinforced concrete, time-domain analysis, wind effects.

Acknowledgements

The author would like to thank H.S. Lew of the National Institute of Standards and Technology for valuable advice and comments, and acknowledges with thanks the previous development of HR_DAD software by Mihai Iancovici, William P. Fritz, René D. Gabbai, and Seymour M.J. Spence during their postdoctoral research associate or guest worker tenures at the National Institute of Standards and Technology. The wind tunnel data developed at the CRIACIV-DIC Boundary Layer Wind Tunnel, Prato, Italy were kindly provided by Dr. Ilaria Venanzi of the University of Perugia. Emil Simiu served as project leader.

Contents

List of Figures

List of Tables

1. Introduction

The ASCE 7-05 Standard (hereinafter ASCE 7-05) (ASCE 2005) specifies three methods for determining wind loads: a simplified, an analytical, and a wind tunnel method. The simplified method is not applicable to flexible, high-rise structures. The analytical method excludes buildings subjected to across-wind and/or aeroelastic effects. Therefore, to the extent that high-rise buildings experience such effects, the analytical procedure is usable only for preliminary design purposes. The wind tunnel method is specified in ASCE 7-05 only in very general terms. This is one of the reasons why estimates of wind effects may vary, in some cases significantly, among independent laboratories or even within the same laboratory (Coffman et al. 2010; Fritz et al. 2008; SOM 2004).

The dynamic response of a building to wind is estimated on the basis of wind tunnel data by using either (1) an approach based on high-frequency force balance (HFFB) measurements of strains at the base of a rigid model or on measurements of strains in an aeroelastic model, or (2) a time-domain approach using simultaneous pressure time-histories on a rigid (Simiu et al. 2008) or aeroelastic model (Diana et al. 2009). The time-domain approach that uses simultaneous time series of pressure data has the following advantages: (1) it does not entail loss of phase information; (2) it can easily account for modes of vibration of any shape, including any number of higher modes of vibration, as well as for mode coupling; and (3) it can account for non-stationary wind effects, provided that corresponding pressure measurements are conducted in specialized test facilities.

The time-domain Database-Assisted Design (DAD) approach, as applied in this report, has been developed with a view to exploiting fully the potential of time-domain approaches (Simiu et al. 2008; Spence 2009). Note the use of the term "Design" in this designation. DAD is not aimed merely at providing the structural engineer with wind loads due to spatially averaged pressures. Rather, DAD is an integrated design methodology that includes member sizing. The member sizing is dictated by (a) the building's aerodynamic and structural properties, (b) the structure's wind environment and its directional interaction with those properties, and (c) design criteria for strength and serviceability. The DAD approach enables the automation of the design of individual structural members of buildings, whether they do or do not experience aeroelastic effects. This study is limited to the case where aeroelastic effects are not significant. Research on the application of DAD to buildings with significant aeroelastic effects is planned for a future study.

The DAD approach allows a clear separation of the wind engineer's and structural engineer's tasks. The wind engineer's task is to produce (1) the required pressure time histories from wind tunnel testing or, as is likely to be the case in the future, from CFD (Computational Fluid Dynamics) simulations, and (2) wind climatological directional data recorded at a weather station reasonably representative of the wind climate at the building site and/or developed by, e.g., Monte Carlo simulations. In addition, it is necessary to produce the ratio between directional wind speeds at the weather station and the reference directional mean wind speeds at the top of the building, given the building's exposure. This ratio enables the transformation of wind tunnel pressure measurements into prototype pressures on the building envelope. Once the wind loading data produced by the wind engineer are available, the structural engineer can use them for quasi-static or dynamic analyses and for accurately determining individual member demand-to-capacity indexes, inter-story drift, and top floor accelerations, corresponding to any specified mean recurrence interval (MRI). The demand-to-capacity index is an indicator of structural strength and adequacy. It incorporates relevant ACI 318 (ACI 2008) and ASCE 7 standard requirements, and is discussed in Chapter 4.

The objective of this paper is to develop and apply the DAD approach to reinforced concrete high-rise buildings. The DAD approach is in our opinion superior from the point of view of physical modeling, accuracy, and convenience to the designer to conventional approaches currently in use. In addition, the approach is transparent, meaning that it can be followed and understood by structural engineers and public officials charged with verifying structural calculations, including their wind engineering components. That wind and structural engineering approaches should satisfy the requirement of transparency would seem obvious. However, as pointed out by SOM (2004), this requirement is not currently met satisfactorily by conventional approaches. The Database-Assisted Design approach has been developed with this requirement in mind.

2. Overview of DAD procedure

The DAD approach as applied to high-rise buildings entails the phases represented in Figure 1. The processes within the dotted box constitute the main algorithm of the High-Rise Database-Assisted Design for Reinforced Concrete structures (HR_DAD_RC) software (NIST 2009). The processes outside the box describe information provided by the wind engineer and the structural engineer. Their tasks are clearly separated in the DAD methodology. The yellow (bright) blocks and the blue (dark) blocks correspond to the wind engineer's and the structural engineer's tasks, respectively. The DAD procedure is described as follows.

1. A preliminary design based on wind speeds specified in the relevant code or specified by the wind engineering consultant is performed by the structural engineer, for example by using the algorithm of ASCE 7-05, Section 6.5. This yields an *initial set of building member dimensions*. The fundamental natural frequencies of vibration for the preliminary design can be obtained by modal analysis using a finite elements analysis (FEA) program. The damping ratios are specified by the structural engineer.

2. Dynamic analyses of the building with the member dimensions determined in phase 1 employ combinations of gravity and wind loads specified in ASCE 7-05, Section 2.3. These combinations can be easily modified if the requirements of the standards change. The analyses are performed by considering the resultant of the wind forces at each floor's mass center, for each wind direction and for reference mean hourly wind speeds at the top of the building of, say, 20 m/s, 30 m/s,…,80 m/s, depending upon the wind speed range of interest at the building location. This phase of the procedure is performed by the structural engineer using as input the directional aerodynamic pressures database provided by the wind engineering consultant. The outputs of this phase are the *floor displacements, floor accelerations, and effective (aerodynamic plus inertial) lateral forces at each floor corresponding to the specified set of directional mean hourly speeds at the top of the building* (e.g., 20 m/s, 30 m/s,.., 80 m/s).

3. The *influence coefficients*, which yield the internal forces in any member due to a unit load with specified direction acting at the mass center of any floor, are calculated by the structural engineer.

4. For each direction and specified wind speed, internal forces induced in members are calculated using the influence coefficients (phase 3) multiplied by the effective floor loads at mass centers (phase 2). The wind-induced forces are added to the respective internal forces induced by factored gravity loads (using the gravity load factor specified in ASCE 7-05, Section 2.3). Demand-to-capacity indexes indicating the extent to which a member is or is not safe are then calculated (see Chapter 4). The output of this phase is a *response database* providing the demand-to-capacity index for the structural members, the inter-story drift along the building height, and the top floor accelerations. The response database is a property of the structure that incorporates its aerodynamic and mechanical characteristics and is independent of the wind climate.

3

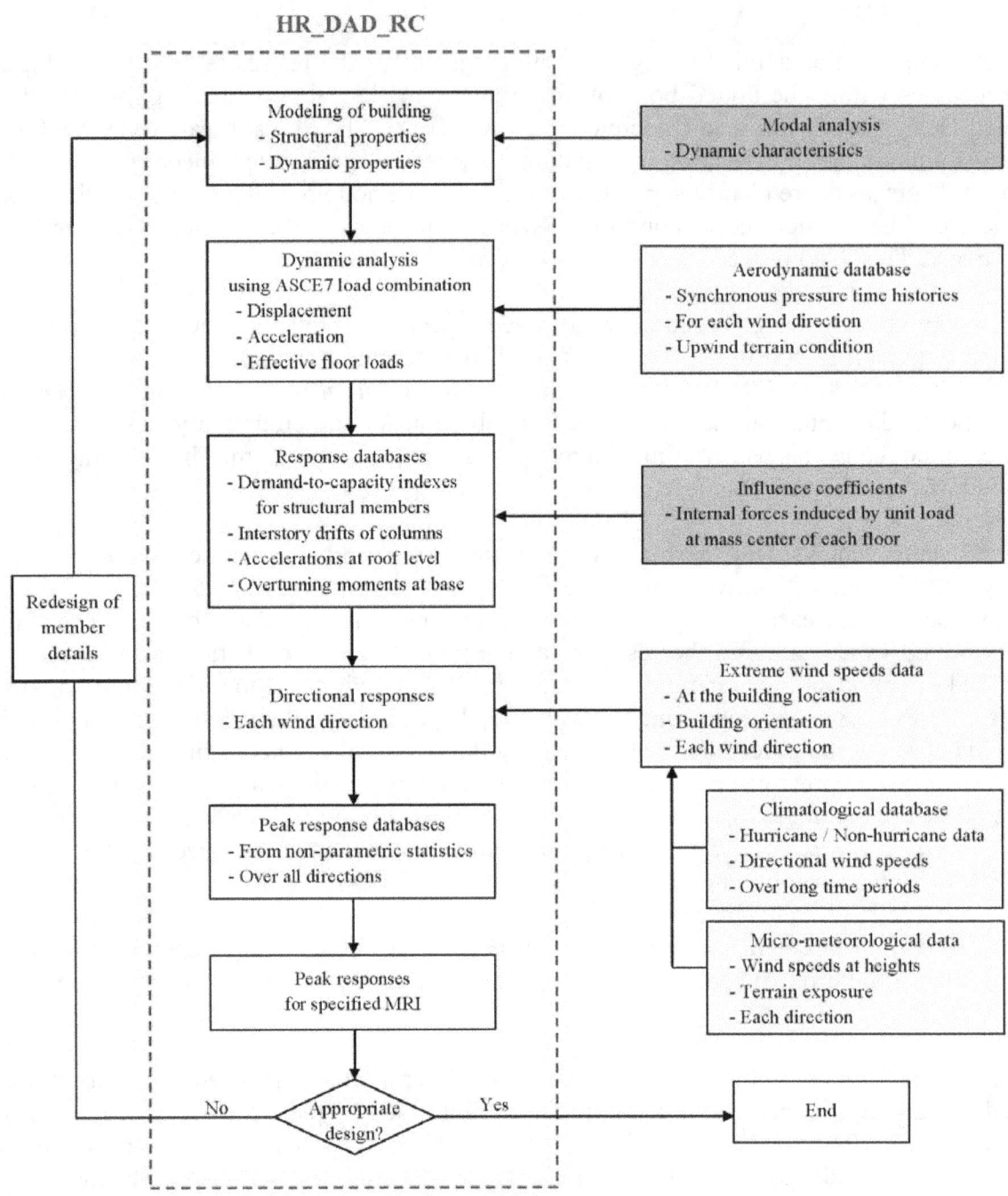

Figure 1. HR_DAD_RC procedure

5. A matrix of directional wind speeds at 10 m above ground in open exposure (i.e., a *wind climatological database*) is developed for a location close to a building of interest. Where necessary, a sufficiently large matrix of wind speeds for each of 36, 16, or 8 directions being considered is developed from measured or simulated wind speed data by using the procedure developed by Grigoriu (2009). Each line of the matrix corresponds to one storm event (if a peaks-over-threshold estimation procedure is used) or to the largest yearly speed (if an epochal estimation procedure is used). The columns of the matrix correspond to the specified wind directions. For hurricane winds, a similar matrix of wind speeds is used. The directional wind speed matrix is provided by the wind engineering consultant. Using micro-meteorological relations, wind tunnel data, or CFD data, the wind engineer also provides a counterpart to this matrix, containing the directional mean hourly wind speeds at the top of the building, in lieu of the directional wind speeds at 10 m above ground in open exposure.

6. Using interpolation procedures, the response database (phase 4) is used in conjunction with the directional wind speed matrix containing the directional hourly mean speeds (phase 5) to calculate a matrix containing the response of interest for each direction of each storm event (or year). However, for each storm event (or year) only the largest of the directional responses is of interest from a design viewpoint and is therefore retained. A one-dimensional vector of the maximum response induced by each storm event is thus created. This vector is then rank-ordered, and the peak responses corresponding to the required mean recurrence intervals are obtained using non-parametric estimation methods (phase 6, see, e.g., Simiu and Miyata, 2006, p. 33). Note that the peak response of interest can consist of the demand-to-capacity indexes for any member, inter-story drift, and peak acceleration for the respective specified MRIs.

7. The procedure outlined in phases 1-6 above is repeated as needed until the results obtained satisfy the design criteria.

3. Analytical framework of DAD

3.1 *Dynamic modeling of building*

A tall building is modeled as a multi-story model with lumped mass and is analyzed by modal analysis. The schematic model of a building is shown in Figure 2. An N-story building has N masses located at the mass center of each floor. The building has three degrees of freedom (i.e., x, y, and θ axes) per floor. The definitions of story and floor used in this report are described in the figure.

The equation of motion for a damped system subject to external loads is given by

$$\mathbf{M}\ddot{\mathbf{u}}(t) + \mathbf{C}\dot{\mathbf{u}}(t) + \mathbf{K}\mathbf{u}(t) = \mathbf{P}(t) \tag{1}$$

where $\mathbf{u}(t)$ is a displacement response vector including displacement in the x and y axes and rotation in the θ axis. $\mathbf{M}$, $\mathbf{C}$, and $\mathbf{K}$, are mass, damping, and stiffness matrices, respectively. $\mathbf{P}(t)$ is a vector of external force time histories.

These coupled equations of motion can be transformed into a set of uncoupled equations by modal analysis. Each uncoupled equation is analogous to the equation of motion for a single degree of freedom system, and can be solved.

By using separation of variables, the total displacement of the structure $\mathbf{u}(t)$ is represented as the summation of its modal contributions, and is dependent on mode shapes $\mathbf{\Phi}$ and generalized coordinates $\mathbf{q}(t)$.

$$\begin{aligned}
\mathbf{u}(t) &= \mathbf{\Phi}\,\mathbf{q}(t) \\
&= \sum_{k=1}^{n} \phi_k\, q_k(t)
\end{aligned} \tag{2}$$

where ϕ_k and q_k are mode shape and generalized coordinate in the kth mode, respectively.

The modal parameters in the modal analysis, including the natural frequency ω, damping ratio ξ, and mode shape $\mathbf{\Phi}$, are the fundamental elements that describe the motion of a structure under forced excitation, and can be obtained from the eigenvalue problem of the equation of motion of the undamped, freely vibrating system.

$$\left[\mathbf{K} - \omega^2 \mathbf{M} \right] \mathbf{\Phi} = 0 \tag{3}$$

The nth mode generalized equation of motion in modal analysis used in DAD is

$$m_n^* \ddot{q}(t) + c_n^* \dot{q}(t) + k_n^* q(t) = p_n^*(t) \tag{4}$$

where the nth mode generalized mass m_n^*, damping c_n^*, stiffness k_n^*, and force p_n^* are defined, respectively, as

6

$$m_n^* = \phi_n^T \mathbf{M} \phi_n$$
$$c_n^* = \phi_n^T \mathbf{C} \phi_n = 2\xi_n m_n^* \omega_n$$
$$k_n^* = \phi_n^T \mathbf{K} \phi_n = m_n^* \omega_n^2 \qquad (5)$$
$$p_n^*(t) = \phi_n^T \mathbf{P}(t)$$

By solving a set of uncoupled equations (Eq. (4)) and accumulating the contribution of their solution to the response, the total displacement, velocity, and acceleration can be obtained (Eq. (2)).

Figure 2 shows a model of N-story building. Each mass has three degrees of freedom (i.e., x, y, and θ axes) and is located on each floor. Definitions of story and floor used in this report are described in the figure. For any given wind direction, wind loads corresponding to the three coordinates are applied at each mass center.

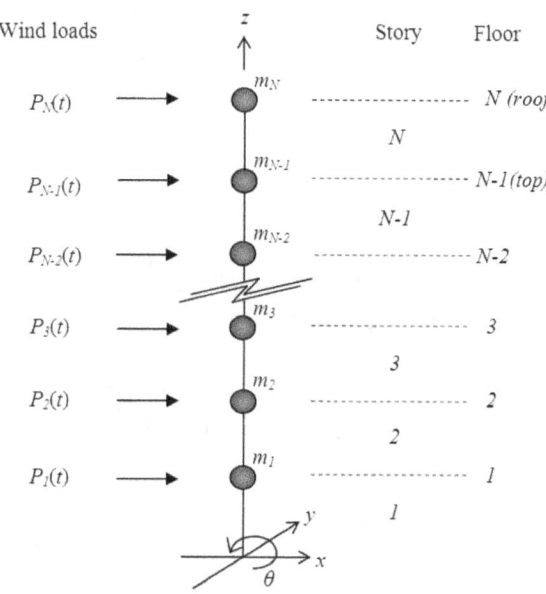

Figure 2. Schematic model of a building

3.2 *Wind load modeling using the aerodynamic database*

DAD employs wind-induced pressure time series on the envelope of the building for each given wind direction. The pressure time histories can be obtained from wind tunnel tests or CFD simulations. This section focuses on load modeling based on wind tunnel tests.

The pressure data on the structure's envelope are generally expressed as non-dimensional pressure coefficient C_p based on the hourly mean wind speed V_H at building roof height H:

$$C_p = \frac{p}{\frac{1}{2}\rho V_H^2} \qquad (6)$$

7

where p is the net pressure relative to the atmospheric pressure. ρ is air density (1.225 kg/m^3) for dry air at sea level under standard atmospheric conditions.

The pressure coefficients may typically be assumed to be identical for the model and the prototype. The pressure time series of the prototype can then be obtained. Prototype wind loads at each mass center along the building height can be calculated using pressures at measuring points and the associated tributary areas (Figure 2).

From the similarity requirement for the reduced frequency (fD/V), where f is sampling frequency and D is the characteristic dimension of the structure, it follows that the time interval Δt_p at prototype scale is

$$\Delta t_p = \frac{D_p}{D_m}\frac{V_m}{V_p}\Delta t_m \tag{7}$$

where D_m/D_p is the length scale, V_m/V_p is the velocity scale, and Δt_m is the time interval (i.e., the reciprocal of the sampling frequency f_m) at model scale. The subscripts p and m stand for prototype and model, respectively. The time interval Δt_p depends on the wind speed V_p.

3.3 Influence coefficients for determining internal forces

The modal equations of motion (Eq. (4)) are solved using the wind loads and the time interval at prototype scale from the previous section. Effective lateral loads $P_e(t)$ acting on each floor mass are

$$P_e(t) = P(t) - m\ddot{u} - c\dot{u} \tag{8}$$

These effective lateral loads, corresponding to each wind direction and speed are combined with dead and live loads using load combination cases specified by ASCE 7-05, Section 2.3:
For strength design,

$$\begin{aligned} 1.2D + 1.0L + 1.0W \qquad &\text{(LC1)} \\ 0.9D + 1.0W \qquad &\text{(LC2)} \end{aligned} \tag{9}$$

For serviceability design,

$$1.0D + 1.0L + 1.0W \tag{10}$$

where D is the total dead load, L is the live load, and W is the wind load. Note that the load factor is not applied to the wind load because the DAD approach yields wind effects with the requisite mean recurrence intervals provided by structural engineer. A different load factor for the live load may be used in Eq. (10) depending upon the engineer's judgment.

These combined loads act on each floor mass in the directions x, y, of the building's principal axes, in the rotation θ, and in the vertical direction z. When these loads are multiplied by appropriate influence coefficients of the structure, they yield the internal forces in any structural member. The influence coefficients consist of the internal forces or moments in any member due to a unit load with specified direction acting at the mass center of any floor. The influence coefficients of a building model can be obtained using a finite element analysis program.

HR_DAD_RC is coded to obtain internal forces at three sections of each structural member. For example, the three locations can be the member's two ends and midpoint. However, in some instances the critical sections, as defined by ACI 318 specifications, may occur at other locations. The influence coefficients required to obtain the internal forces must be calculated accordingly.

3.4 *Response database*

DAD calculates response databases for a sufficient number of wind directions and speeds. The databases contain information on structural responses induced by combined gravity and lateral wind loads. The structural responses used in DAD are demand-to-capacity indexes, inter-story drift, and peak accelerations. Details are provided in Chapter 4.

For the design of structural members, response databases pertaining to demand-to-capacity indexes are obtained for individual structural members (e.g., columns and beams) for the wind directions and speeds of interest. The index " B_{ij}^{PM} " for columns pertains to interaction equations for axial load and bending moments; for beams the index is based on the bending moment. The index " B_{ij}^{VT} " for columns and beams is associated with interaction equations for shear forces and torsional moment.

For serviceability design, response databases for inter-story drift are obtained along specified column lines, and response databases of peak accelerations are obtained at the corners of the top floor. If higher modes participate in the response the peak accelerations may occur at floors other than the top floor.

The wind directions θ_w in the response databases include directions in increments of 10 °, i.e., 0 °, 10 °, 20 °, ... , 360 ° (Figure 3). The wind speeds include all speeds in the range of interest in increments of, say, 10 m/s, e.g., from 20 m/s to 80 m/s. Using interpolation techniques the structural response can be obtained for any wind direction and speed within the specified ranges. As noted earlier the directions for which the responses are provided in the response databases are referenced to the principal axes of the building (i.e., the x and y axes in Figure 3). The response databases are independent of wind climate.

3.5 *Peak directional response*

Structural responses to directional wind are obtained by combining the directional wind speeds of the wind climatological database with the response databases. The wind climatological database for the calculations presented in this report is based on a wind speed dataset of 999 simulated hurricanes for 16 directions (Batts et al. 1980). The wind climatological database is publicly available at www.nist.gov/wind.

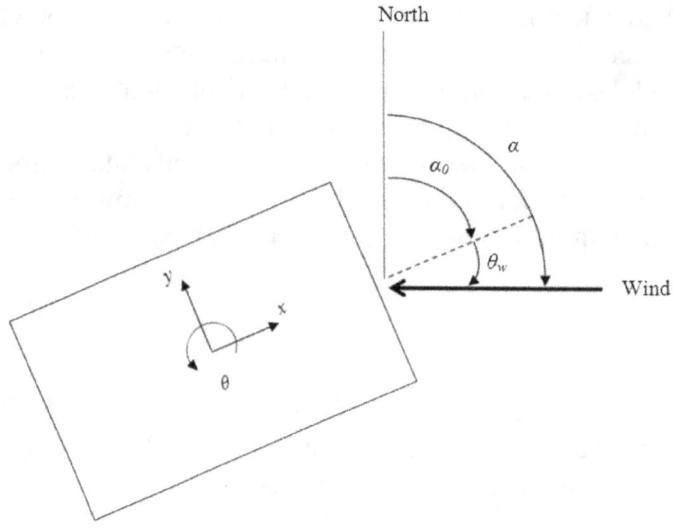

Figure 3. Wind directions

The database sets in the NIST website consist of one-minute mean hurricane wind speeds in knots at 10 m above the ground in open terrain near the coast line. The 16 wind directions are from 22.5° to 360° in 22.5° clockwise increments from the North. Those speeds are converted to hourly mean speeds (m/s) at the rooftop of the building corresponding to the terrain exposure at the site. Providing the original and converted wind climatological database is incumbent on the wind engineering consultant.

DAD also accounts for building orientation, defined by the angle α_0 in Figure 3. Because the wind direction θ_w angle in the response database is not identical to the wind direction angle α in the climatological database (Figure 3), each wind direction in the wind climatological database should be rotated by the orientation angle α_0.

For each storm event DAD retains only the largest of the directional responses induced the wind speeds associated with that event. A one-dimensional vector of the largest responses so obtained, whose dimension is equal to the number of storm events in the wind climatological database, is thus produced. The peak responses corresponding to the mean recurrence intervals (MRIs) of interest are then calculated from this vector using a non-parametric estimation method presented in Section 3.7. It is again emphasized that the estimated peak responses with specified mean recurrence intervals are obtained for wind load effects, not for wind loads.

3.6 Veering effects

The DAD methodology can account for veering as a function of height above ground, surface exposure, and angle of latitude of the building's location. The response database is calculated from the aerodynamic database whose reference wind speed is measured at the top of the building. On the other hand, before conversion to the height of the top of the building, the wind climatological database is referenced to speeds at the standard elevation (typically 10 m). The conversion of the wind speeds at 10 m to mean hourly speeds at the height of the top of the building must therefore be accompanied by a rotation (Yeo and Simiu, (2010).

3.7 Peak wind effects with specified MRIs

The time series of peak wind effects induced by each storm event in the wind climatological database is used to obtain the requisite peak wind effect with the specified MRI. The time series is rank ordered, the largest wind effect having rank one, and the non-parametric estimation method described in Section 2.4.3.2.2 of Simiu and Miyata (2006) can be employed.

Based on the assumption that the occurrence of storm events is a Poisson process with constant occurrence rate, the estimated MRI $\overline{N_k}$ associated with k^{th} ranked peak wind effects is

$$\overline{N_k} = \frac{n+1}{\nu k} \tag{11}$$

where n is total number of storm events in the database. Interpolation is used where necessary.

3.8 Adjustment of demand-to-capacity indexes

According to ASCE 7-05, Section C6.6, it is prudent for estimates based on the wind tunnel method to be not less than 80 % of the corresponding estimates based on the ASCE 7 analytical method. For practical reasons this requirement applies to estimates of peak overturning moments in the principal axes with MRIs specified in the Standard.

Calculations of overturning moments based on ASCE 7-05 are performed by using the importance factor 0.87 (Risk Category I buildings), 1.0 (Risk Category II buildings), and 1.15 (Risk Category III or IV buildings) (Table 6-1, ASCE 7-05). Twenty-four load cases using equivalent static wind loads are employed in accordance with combinations of wind loads defined in Figure 6-9 of ASCE 7-05.

The peak overturning moments determined by the DAD procedure for the MRI corresponding to appropriate ASCE 7 Risk Category are estimated by using response and climatological databases. If the moments in DAD are less than 80 % of those determined in accordance with Section 6.5 of ASCE 7-05, the demand-to-capacity index is adjusted as follows:

$$B_{ij}^* = \gamma B_{ij}$$
$$\gamma = \frac{0.8}{M_o^{DAD} / M_o^{ASCE7}} \tag{12}$$

where M_o^{DAD} and M_o^{ASCE7} are the overturning moments at base obtained from DAD and Section 6.5, ASCE 7-05, respectively, and γ is the index adjustment factor. If the moment in DAD is not less than 80 % of the ASCE 7-05 value, the index need not be modified (i.e., $B_{ij}^* = B_{ij}$).

3.9 Compliance with design criteria

Once peak responses (i.e., demand-to-capacity index, inter-story drift, and acceleration) for specified MRIs are obtained, DAD verifies if the peak responses satisfy design criteria for safety and serviceability.

Peak demand-to-capacity indexes accounting for ASCE 7 limitations on overturning moments must be equal to or less than unity for any structural member with an appropriate MRI, i.e.,

$B_{ij}^{PM*} \leq 1$ and $B_{ij}^{VT*} \leq 1$. If these relations are satisfied, the structural members are adequately designed for axial force and bending moments, and for shear force and torsional moments, respectively.

The peak inter-story drift ratio (defined as the ratio of drift to story height, see Eq. (18)) is not limited by ASCE 7-05 requirements. However, the ASCE 7-05 Commentary suggests limits on the order of 1/600 to 1/400 (see Section CC.1.2 in ASCE 7-05). If the peak inter-story drift ratio with a 20-year MRI is equal to or less than the suggested limit, the design is adequate for inter-story drift. This criterion may be changed as judged appropriate by the building owner and designer.

Peak resultant accelerations at the top floor are also not limited by ASCE 7-05. This study assumes a limit of 25 mg for a 10-year MRI for office buildings (Isyumov et al. 1992). If peak acceleration for MRI = 10 year is not greater than this limit the design is adequate for peak acceleration.

The procedure outlined in Sections 3.1 to 3.8 is repeated as needed with a modified structural design (e.g., by resizing members or by installing dampers) until the results satisfy the design criteria.

4. Structural responses considered in design

The loads induce in the structure three types of response considered in design: demand-to capacity indexes (Sect. 4.1), inter-story drift (Sect. 4.2), and accelerations at the top floor (Sect. 4.3).

4.1 *Demand-to-capacity indexes*

A demand-to-capacity index is a quantity used to measure the adequacy of a structural member's strength. In general, this index is defined as a ratio or sum of ratios of the internal force induced by design loads to associated strength provided by the section. The design strength in HR_DAD_RC is based on the Building Code Requirements for Structural Concrete and Commentary 318-08 (hereinafter ACI 318-08). An index higher than unity indicates inadequate design of a structural member; the index must be less or equal to unity for the design satisfaction. The HR_DAD_RC software employs two demand-to-capacity indexes: (1) for axial and/or flexural loads, and (2) for shear and torsion.

The index " B_{ij}^{PM} " pertains to interaction for axial load and bending moments for columns and bending moments for beams:

$$
\begin{aligned}
B_{ij}^{PM} &= \frac{M_u}{\phi_m M_n} \qquad \text{(for a tension-controlled section)} \\
&= \frac{P_u}{\phi_p P_n} \qquad \text{(for a compression-controlled section)}
\end{aligned}
\tag{13}
$$

where M_u and P_u are the factored bending moment and axial force at the section, M_n and P_n are the nominal moment and axial strengths at the section, and ϕ_m and ϕ_p are the reduction factors for flexural and axial strengths, respectively.

For columns subject to bi-axial flexure loads, the PCA (Portland Cement Assosication) load contour method (PCA 2008) is used for tension-controlled sections:

$$
\begin{aligned}
B_{ij}^{PM} &= \frac{M_{ux}}{\phi M_{nox}}\left(\frac{1-\beta}{\beta}\right) + \frac{M_{uy}}{\phi M_{noy}} \qquad for \quad \frac{M_{uy}}{M_{ux}} > \frac{M_{noy}}{M_{nox}} \\
&= \frac{M_{ux}}{\phi M_{nox}} + \frac{M_{uy}}{\phi M_{noy}}\left(\frac{1-\beta}{\beta}\right) \qquad for \quad \frac{M_{uy}}{M_{ux}} < \frac{M_{noy}}{M_{nox}}
\end{aligned}
\tag{14}
$$

where M_{ux} is the factored bending moment about x-axis, M_{uy} is the factored bending moment about y-axis, M_{nox} is the nominal uniaxial moment strength about x-axis, M_{noy} is the nominal uniaxial moment strength about y-axis, and β is a constant in relation to properties and details of the member (0.65 is used as approximation).

The Bresler reciprocal load method in ACI 318-08 (R10.3.6) is used for compression-controlled sections:

$$B_{ij}^{PM} = \frac{P_u}{\phi P_n}$$

$$= \frac{P_u}{\phi \dfrac{1}{\dfrac{1}{P_{ox}} + \dfrac{1}{P_{oy}} - \dfrac{1}{P_o}}} \tag{15}$$

where P_{ox} is the maximum uniaxial load strength of the column with moment M_{nx} ($= P_n\, e_y$; e_y is the eccentricity along y-axis), P_{oy} is the maximum uniaxial load strength of the column with moment M_{ny} ($= P_n\, e_x$; e_x is the eccentricity along x-axis), and P_o is the maximum axial load strength with no applied moments.

The index " B_{ij}^{VT} " is associated with interaction equations for shear forces and torsional moment for columns and beams:

$$B_{ij}^{VT} = \frac{\sqrt{V_u^2 + \left(\dfrac{T_u p_h b_w d}{1.7 A_{oh}^2}\right)^2}}{\phi_v \left(V_c + V_s\right)} \tag{16}$$

where V_c and V_s are the nominal shear strengths provided by concrete and by reinforcement, respectively, V_u is the shear force, T_u is the torsional moment, ϕ_v is the reduction factors for shear strengths, p_h is the perimeter enclosed by the centerline of the outermost closed stirrups, A_{oh} is the area enclosed by the centerline of the outermost closed stirrups, b_w is the width of the member, and d is the distance from extreme compression fiber to the centroid of longitudinal tension reinforcement.

For members subject to bi-axial shear forces, the index is

$$B_{ij}^{VT} = \frac{\sqrt{V_{ux}^2 + V_{uy}^2 + \left(\dfrac{T_u p_h b_w d}{1.7 A_{oh}^2}\right)^2}}{\phi_v \left(V_c + V_s\right)} \tag{17}$$

where V_{ux} and V_{uy} are the shear forces in the x and y axes, respectively.

4.2 Inter-story drift

The time series of the inter-story drift ratios at the ith story, $d_{i,x}(t)$ and $d_{i,y}(t)$, corresponding to the x and y axes, are:

$$d_{i,x}(t) = \frac{\left[x_i(t) - D_{i,y}\theta_i(t)\right] - \left[x_{i-1}(t) - D_{i-1,y}\theta_{i-1}(t)\right]}{h_i}$$

$$d_{i,y}(t) = \frac{\left[y_i(t) + D_{i,x}\theta_i(t)\right] - \left[y_{i-1}(t) + D_{i-1,x}\theta_{i-1}(t)\right]}{h_i}$$

(18)

where $x_i(t)$, $y_i(t)$, and $\theta_i(t)$ are the displacements and rotation at the mass center at the ith floor, $D_{i,x}$ and $D_{i,y}$ are distances along the x and y axes from the mass center on the ith floor to the point of interest on that floor (Figure 4), and h_i is the ith story height between mass centers of the ith and the i-1th floor.

The ASCE 7-05 Commentary (2005) suggests limits on the order of 1/600 to 1/400 (see Section CC.1.2 in ASCE 7-05).

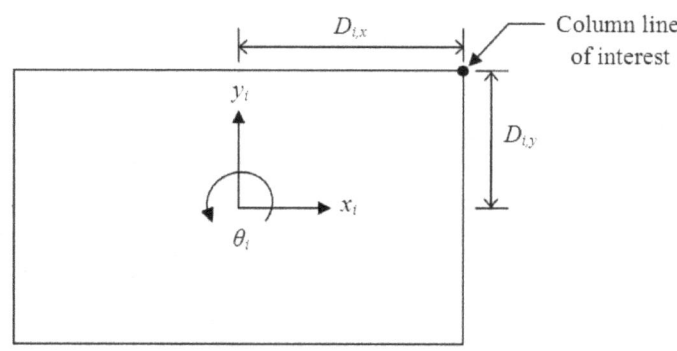

Figure 4. Position parameters at floor i for inter-story drift

4.3 *Top floor acceleration*

The time series of resultant acceleration at the top floor, $a_r(t)$ are yielded in HR_DAD_RC by the expression:

$$a_r(t) = \sqrt{\left[\ddot{x}_{top}(t) - D_{top,y}\ddot{\theta}_{top}(t)\right]^2 + \left[\ddot{y}_{top}(t) + D_{top,x}\ddot{\theta}_{top}(t)\right]^2}$$

(19)

where accelerations $\ddot{x}_{top}(t)$, $\ddot{y}_{top}(t)$, and $\ddot{\theta}_{top}(t)$ of the mass center at the top floor pertain to the x, y, and θ (i.e., rotational) axes, and $D_{top,x}$ and $D_{top,y}$ are the distances along the x and y axes from the mass center to the point of interest on the top floor.

The resultant value of Eq. (19) is used, rather than accelerations along the principal axes, because peak acceleration is of concern for human discomfort regardless of its direction. While ASCE 7-05 does not provide wind-related peak acceleration limits, for office buildings a limit of 25 mg with a 10-year MRI was suggested by Isyumov et al. (1992) and Kareem et al. (1999).

5. Application to a 60-story CAARC building

A 60-story reinforced concrete building with rigid diaphragm floors (Figure 5) was designed using the High-Rise Database-Assisted Design for Reinforced Concrete structures (HR_DAD_RC) software. The dimensions of the building are 45.72 m width (dimension B in Figure 5), 30.48 m depth (D), and 182.88 m height (H), and define the Commonwealth Advisory Aeronautical Research Council (CAARC) building studied by various researchers (Melbourne 1980; Venanzi 2005; Wardlaw and Moss 1971). The building has a moment-resisting frame structural system similar to the structural system studied by Teshigawara (2001), and consists of 2880 columns and 4920 beams, in addition to rigid diaphragm slabs. The building was assumed to be located near Miami, Florida and to have suburban exposure.

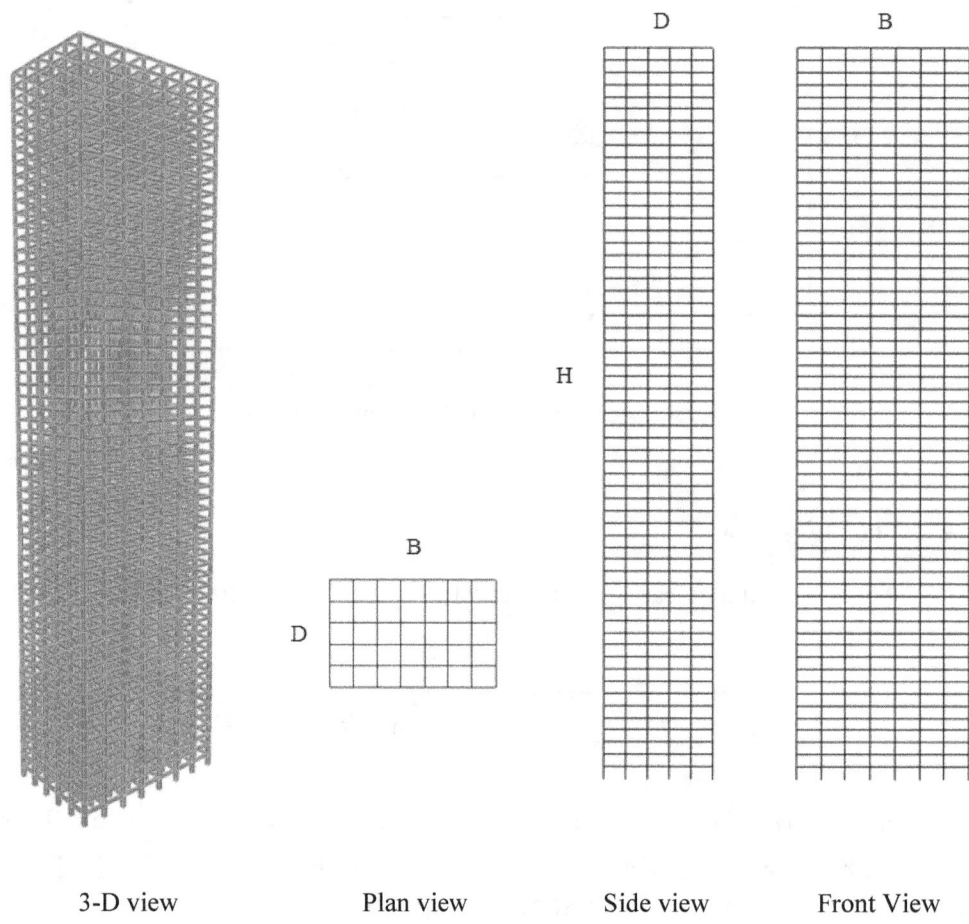

| 3-D view | Plan view | Side view | Front View |

Figure 5. Schematic views of a 60-story building

5.1 *Modeling of a building*

The building was first designed by using the algorithm in Section 6.5, ASCE 7-05. Once initial dimensions of members in the building were obtained, natural frequencies of vibration and mode shapes were calculated by modal analysis using a finite elements analysis program. The modal damping ratios were assumed to be 2 % in all three modes considered in this study (Table 1).

As shown in Figures 5 and 6, structural members of the building consist of columns, beams, and slabs. (The design of the slabs was not performed in this study.) Columns were categorized as corner and non-corner columns. Beam members were divided into exterior (spandrel) and interior beams. The building is comprised of six sets of members. Each set consists of 10 stories for which the member dimensions are the same. The first set applies to the first ten stories, the second to the next ten stories, and so forth. The compressive strengths of concrete for all members are 80 MPa from the 1st to the 40th story and 60 MPa from the 41st to 60th story. The dimensions and reinforcement details are listed in Table 2. Columns have longitudinal reinforcement uniformly distributed along the sides and hoops, and beams have tensile and compression reinforcement and stirrups. The yield strengths of reinforcements are 520 MPa for longitudinal bars and 420 MPa for hoop or stirrup bars. Wind effects were calculated for a typical set of 96 members (Table 2).

Table 1. Dynamic properties of a building

Mode	1st (*y* dir.)	2nd (*x* dir.)	3rd (θ dir.)
Natural frequency [Hz]	0.165	0.175	0.200
Damping ratio [%]	2.0	2.0	2.0

5.2 *Dynamic analysis using aerodynamic database*

Time histories of aerodynamic wind loads on each floor were calculated from time-series of pressure on a rigid wind tunnel model of the CAARC building, measured for winds with directions with 10° increments for suburban terrain exposure. The wind tunnel tests were performed at the Inter-University Research Center on Building Aerodynamics and Wind Engineering (CRIA-CIV-DIC) Boundary Layer Wind Tunnel in Prato, Italy (Venanzi 2005). Note that for buildings sensitive to aeroelastic phenomena, synchronous pressures must also be measured on an aeroelastic model under a range of wind speeds and directions for which aeroelastic responses occur (Diana et al. 2009). However, in this study aeroelastic effects are assumed not to be present.

Dynamic analyses of the building were performed by using wind loads corresponding to wind speeds of 20 m/s to 80 m/s in increments of 10 m/s, using the directional pressure data obtained from the wind tunnel tests for suburban terrain exposure. Time series of displacements and accelerations at the mass center on each floor were obtained, and effective lateral loads on all floors were calculated for each wind speed and wind direction.

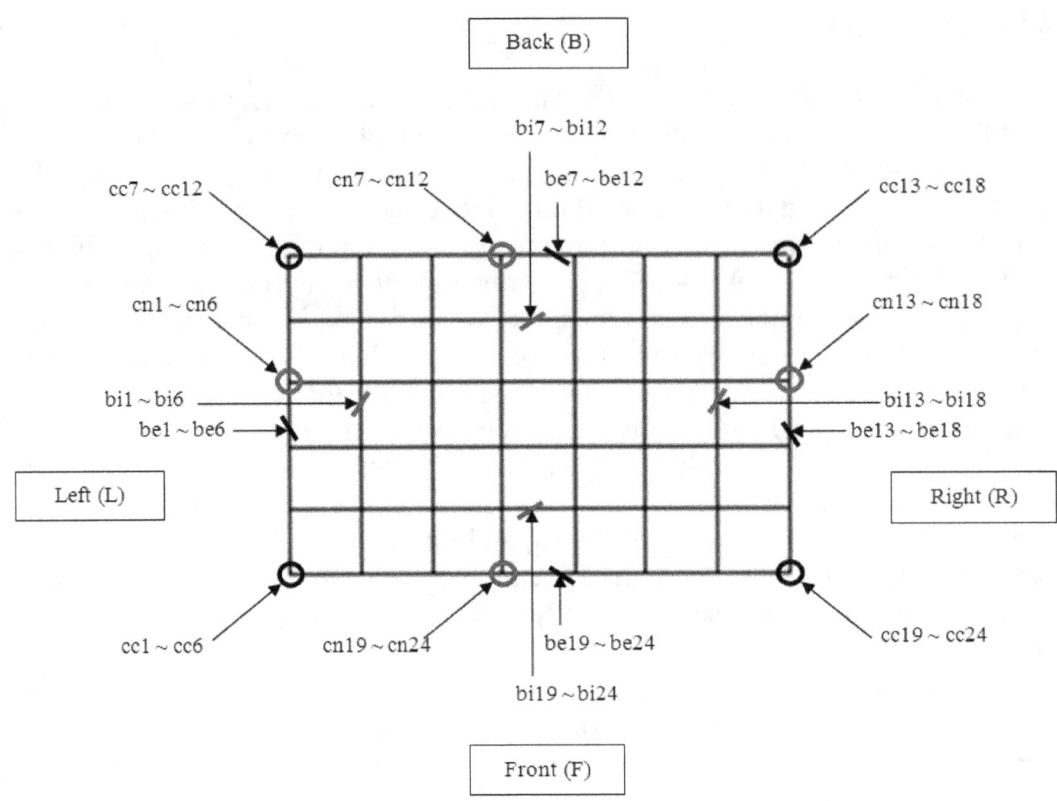

Figure 6. Plan view of building with locations of selected members ($\alpha_0 = 90°$)

(cc = corner column; cn = non-corner column; be = exterior beam; bi = interior beam)

Two load combinations, LC1 and LC2, associated with wind and gravity loads for each wind direction and speed (Eq. (9)) were considered for strength design. One case (Eq. (10)) was considered for serviceability design.

5.3 *Response database*

Response databases for demand-to-capacity indexes, inter-story drift, and peak acceleration, were calculated for the wind directions and speeds being considered in each load combination case. For strength design, the response database for the demand-to-capacity indexes (Eqs. (13) and (16)) were calculated for each individual member. For serviceability design, response databases for inter-story drift and acceleration (Eqs. (18) and (19)) were calculated for a column line and a top floor corner, respectively.

Response databases of demand-to-capacity indexes are shown for a corner column (cc) under LC2 in Figure 7 where θ_w is the wind direction (Figure 3). Figures 8 and 9 show response databases for *y*-axis inter-story drifts on the 43rd story and the response database for the resultant acceleration on the top floor of the corner at the intersection of the front and left sides (Figure 6), respectively.

Table 2. Section dimensions and reinforcement details of structural members

Name	Story	Section [mm × mm]	Longitudinal bar (Longitudinal)	Hoop or stirrup [spacing: mm]	Selected member
Corner column (cc)	$51^{st}\sim60^{th}$	750 × 750	12 - D29	4 - D13@200	6, 12, 18, 24 (51st st.)
	$41^{st}\sim50^{th}$	750 × 750	12 - D29	4 - D13@200	5, 11, 17, 23 (41st st.)
	$31^{st}\sim40^{th}$	800 × 800	16 - D32	4 - D13@200	4, 10, 16, 22 (31st st.)
	$21^{st}\sim30^{th}$	850 × 850	20 - D32	4 - D16@200	3, 9, 15, 21 (21st st.)
	$11^{th}\sim20^{th}$	900 × 900	20+12 - D43	4 - D16@200	2, 8, 14, 20 (11th st.)
	$1^{st}\sim10^{th}$	1100 × 1100	24+16 - D43	4 - D16@200	1, 7, 13, 19 (1st st.)
Non-corner column (cn)	$51^{st}\sim60^{th}$	750 × 750	12 - D25	4 - D13@200	6, 12, 18, 24 (51st st.)
	$41^{st}\sim50^{th}$	750 × 750	12 - D25	4 - D13@200	5, 11, 17, 23 (41st st.)
	$31^{st}\sim40^{th}$	800 × 800	12 - D25	4 - D16@200	4, 10, 16, 22 (31st st.)
	$21^{st}\sim30^{th}$	850 × 850	16 - D29	4 - D16@200	3, 9, 15, 21 (21st st.)
	$11^{th}\sim20^{th}$	900 × 900	20+12 - D43	4 - D16@200	2, 8, 14, 20 (11th st.)
	$1^{st}\sim10^{th}$	1100 × 1100	20+16 - D43	4 - D16@200	1, 7, 13, 19 (1st st.)
Exterior beam (be)	$51^{st}\sim60^{th}$	400 × 700	4 - D32 / 2 - D32	2 - D13@150	6, 12, 18, 24 (roof)
	$41^{st}\sim50^{th}$	400 × 700	4+4 - D32 / 3 - D32	2 - D16@150	5, 11, 17, 23 (50th fl.)
	$31^{st}\sim40^{th}$	450 × 750	4+4 - D36 / 4 - D32	4 - D16@150	4, 10, 16, 22 (40th fl.)
	$21^{st}\sim30^{th}$	500 × 750	5+5 - D36 / 4 - D36	4 - D16@150	3, 9, 15, 21 (30th fl.)
	$11^{th}\sim20^{th}$	550 × 750	5+5 - D43 / 4 - D36	4 - D16@150	2, 8, 14, 20 (20th fl.)
	$1^{st}\sim10^{th}$	550 × 800	5+5 - D43 / 4 - D36	4 - D16@150	1, 7, 13, 19 (10th fl.)
Interior beam (bi)	$51^{st}\sim60^{th}$	400 × 700	4 - D29 / 2 - D29	2 - D13@150	6, 12, 18, 24 (roof)
	$41^{st}\sim50^{th}$	400 × 700	4+4 - D32 / 2 - D32	2 - D13@150	5, 11, 17, 23 (50th fl.)
	$31^{st}\sim40^{th}$	450 × 750	4+4 - D36 / 3 - D32	4 - D13@150	4, 10, 16, 22 (40th fl.)
	$21^{st}\sim30^{th}$	500 × 750	5+5 - D36 / 4 - D36	4 - D13@150	3, 9, 15, 21 (30th fl.)
	$11^{th}\sim20^{th}$	550 × 750	5+5 - D36 / 4 - D36	4 - D13@150	2, 8, 14, 20 (20th fl.)
	$1^{st}\sim10^{th}$	550 × 800	5+5 - D36 / 4 - D36	4 - D13@150	1, 7, 13, 19 (10th fl.)

Note: cc1~cc24 for corner columns; cn1~cn24 for non-corner columns; be1~be24 for exterior beams; bi1~bi24 for interior beams; st. for story; fl. for floor.

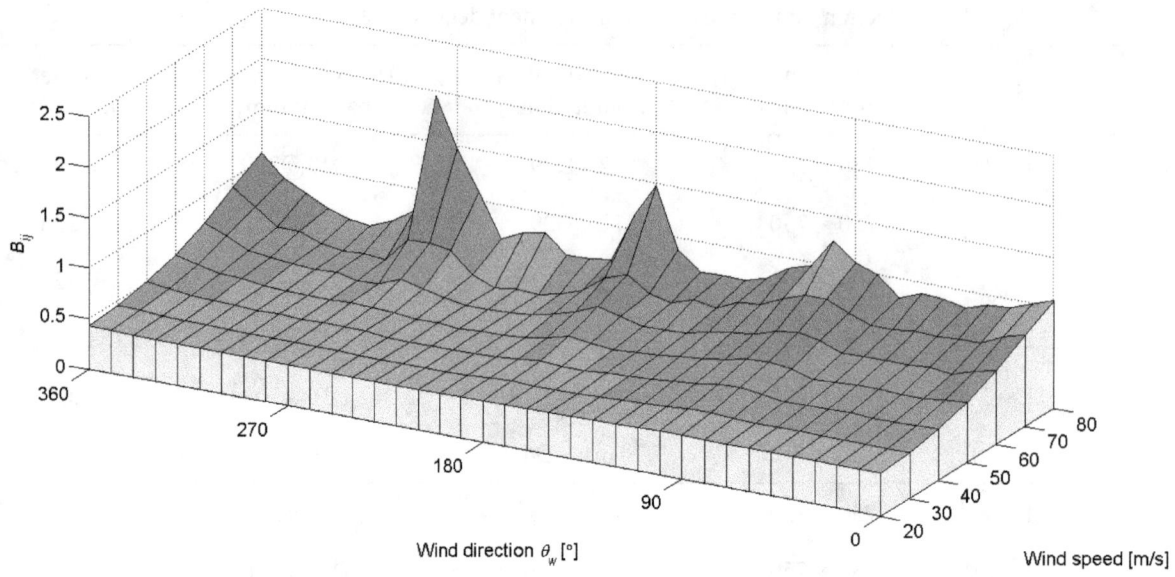

(a) B_{ij}^{PM}

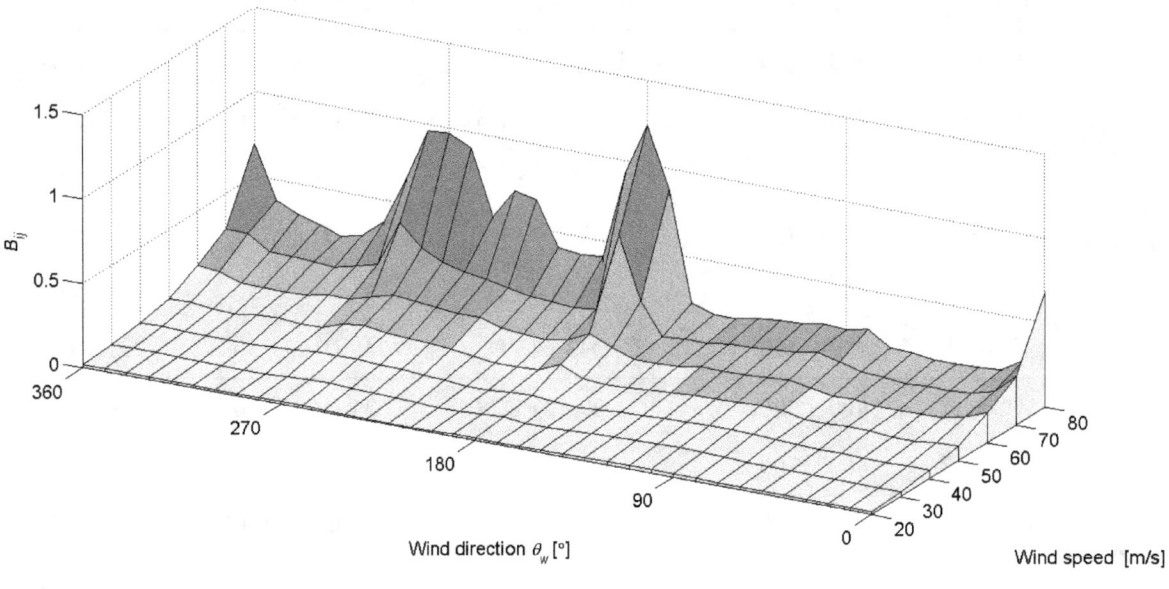

(b) B_{ij}^{VT}

Figure 7. Response database: demand-to-capacity index (member ID = cc7)

20

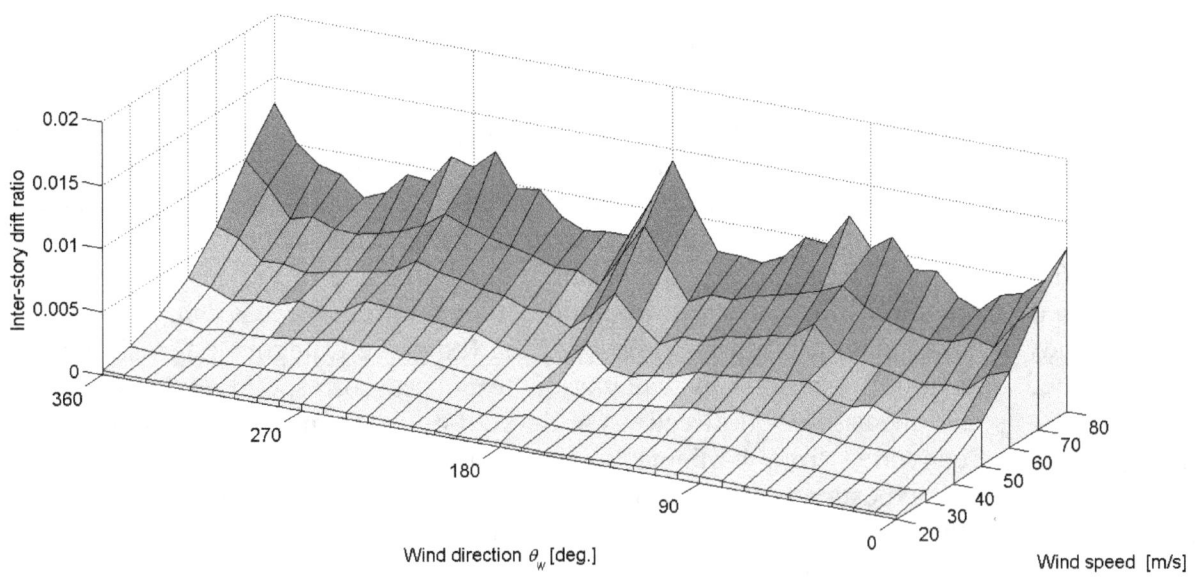

Figure 8. Response database: *y*-axis inter-story drift

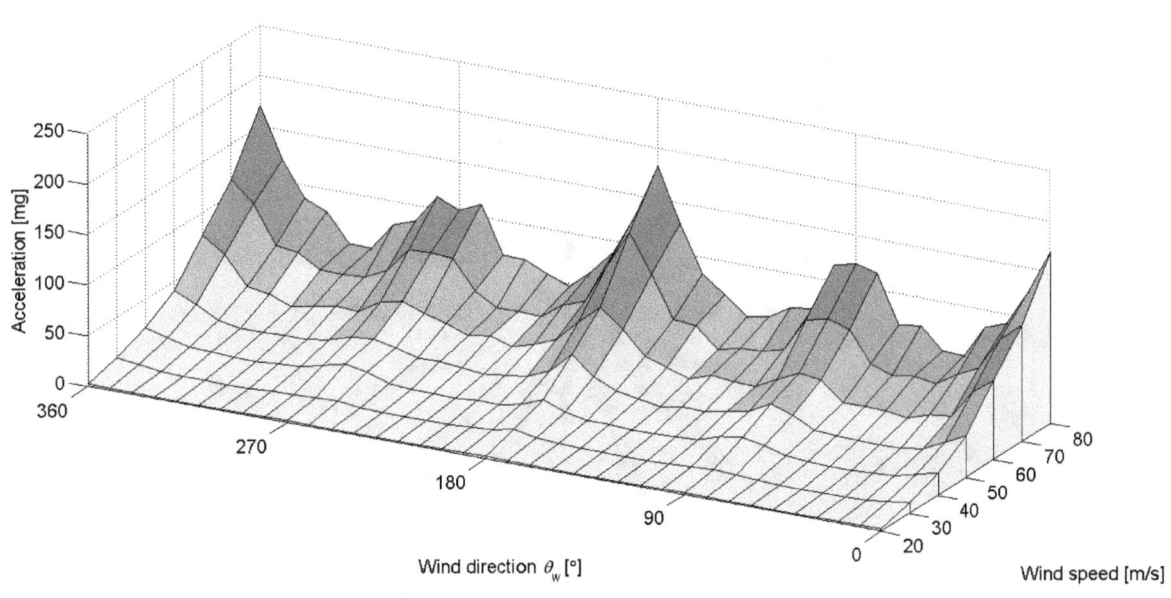

Figure 9. Response database: resultant acceleration

21

5.4 *Directional responses*

Structural responses under wind at the building location were obtained by applying to the response databases the directional wind speeds from the climatological database at a building location. The climatological database used in the study is a dataset of 999 simulated hurricanes with wind speeds for 16 directions near Miami, Florida (Milepost 1450), and was obtained from www.nist.gov/wind. The angles α indicating those directions are from 22.5° to 360° clockwise from the North in 22.5° increments. Figure 10 shows directional hurricane wind speeds according to the 16 directions at Miami. In this study, the orientation angle of the design building is α_0 = 90° clockwise from the North, that is, the front side of the building faces South.

This study assumed suburban terrain exposure (i.e., Exposure Category B) in all directions. DAD obtained directional responses by calculating responses corresponding to hourly mean wind speeds (m/s) and the associated directions at the rooftop in suburban terrain exposure. The wind speeds were converted from 1-minute hurricane wind speeds (in knots) at 10 m above ground in open terrain exposure to hourly wind speeds at the elevation of the top of the building, and then applied to the response databases, given the building orientation α_0. Veering effects (see Yeo and Simiu (2010) were not considered in this study.

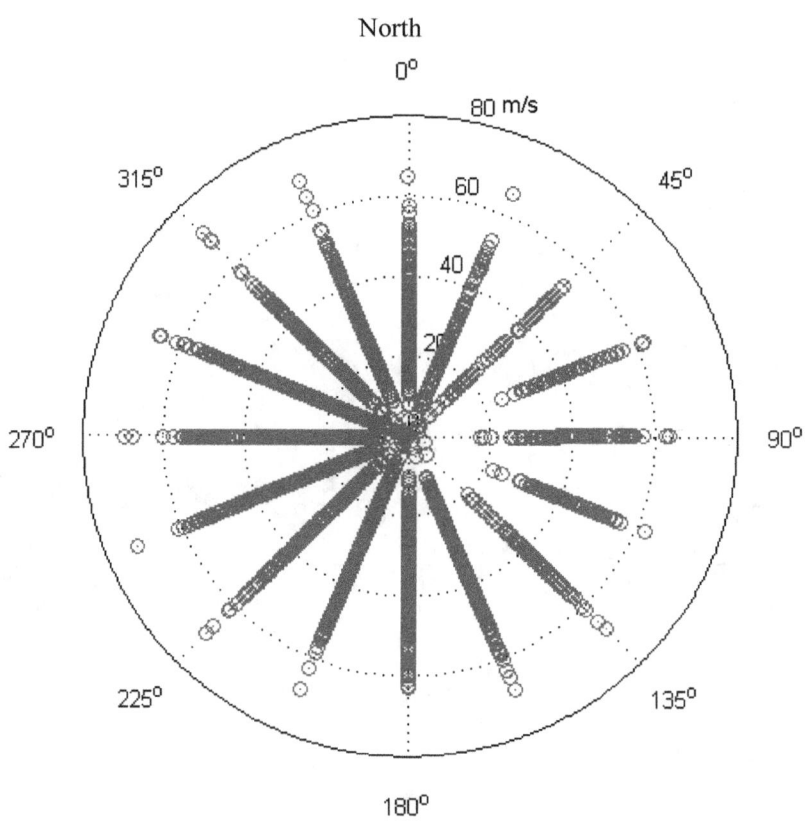

Figure 10. Hurricane wind speeds (Miami)

The peak response database consists in each case of the respective vector of the 999 largest responses (Eq. (11)). Examples of peak response databases for LC2 are shown for demand-to-capacity indexes of a corner column of cc7 (Figure 11), inter-story drift of the front-left corner at the 43rd story (Figure 12), and peak accelerations of the front-left corner of the top floor (Figure 13). The figures show that the peak responses increase monotonically with MRI. Note that the peak responses of inter-story drifts and accelerations along both principal axes do not occur at the same time.

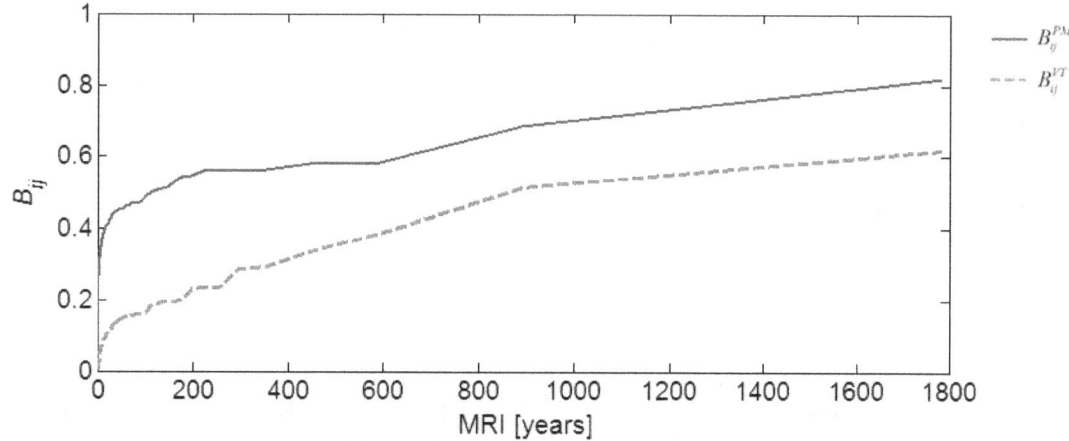

Figure 11. Peak response database: demand-to-capacity index in Load Case 2 (LC2)

(member ID = cc7)

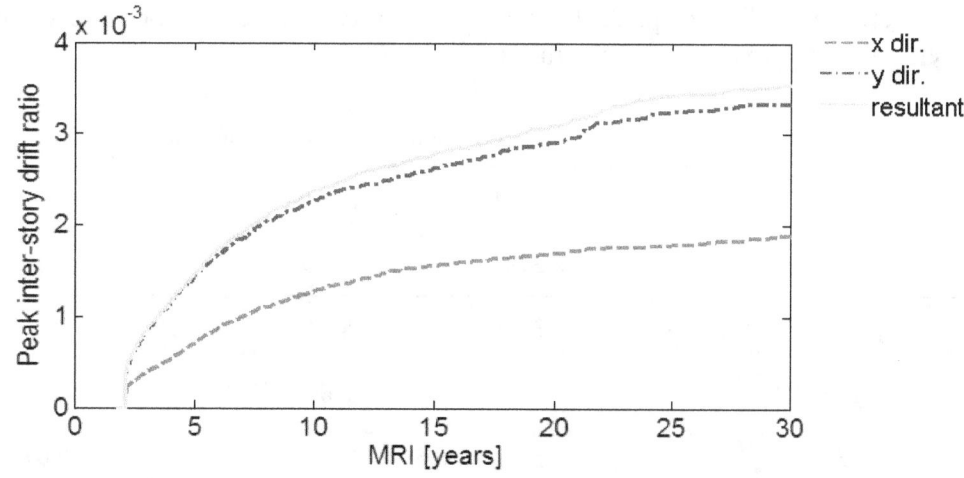

Figure 12. Peak response database: inter-story drift
(the front-left corner at the 43rd story)

23

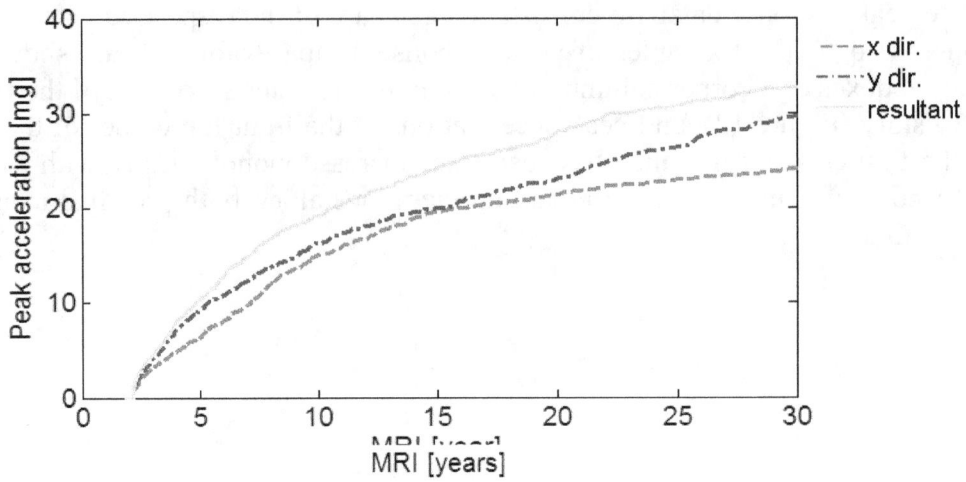

Figure 13. Peak response database: acceleration
(the front-left corner of the top floor)

5.5 *Adjustment of demand-to-capacity indexes*

As an option, DAD accounts for the ASCE 7-05 design requirement that forces and pressures estimated through wind tunnel testing are to be limited to not less than 80% of its ASCE 7-based counterpart (see ASCE 7-05 Section C6.6). This study calculated ASCE 7-based overturning moments in the principal axes (i.e., x and y axes) of buildings with Risk Category "II" and "III or IV" and compared them to the peak overturning moments determined by the DAD procedure for MRI = 700 years and 1700 years, respectively.

As shown in Table 3, ratios of overturning moments from DAD to those from ASCE 7 are less than 0.8 for both MRIs; the corresponding index adjustment factors γ (Eq. (12)) are 1.16 and 1.19, respectively. Adjusted peak demand-to-capacity indexes for both MRIs were obtained by multiplying the indexes by the adjustment factors.

Table 3. Overturning moments and adjustment factor

	MRI = 700 years		MRI = 1700 years	
	ASCE 7	DAD	ASCE 7	DAD
M_{ox} [×10^6 kN·m]	6.10	4.22	7.01	4.70
M_{oy} [×10^6 kN·m]	3.36	2.64	3.87	2.86
$M_{ox}^{DAD}/M_{ox}^{ASCE7}$	0.69		0.67	
$M_{oy}^{DAD}/M_{oy}^{ASCE7}$	0.79		0.74	
γ	1.16		1.19	

24

Peak responses were obtained for adjusted demand-to-capacity indexes corresponding to the MRIs of 700 and 1700 years. Maximum values of peak adjusted demand-to-capacity indexes for the 96 selected members are summarized in Table 4 as functions of the load combination being considered. For the index B_{ij}^{PM*} the load combination case of LC1 governs. Significant differences between the LC1 and LC2 cases occur for the columns. In particular, the index for corner columns changes noticeably if the MRI increases from 700 to 1700 years. This is due to the fact that lower axial compression forces reduce the flexural strength of a column with a tension-controlled section. For the index B_{ij}^{VT*} LC2 governs for columns and LC1 for beams. It is notable that the increases in both indexes are generally larger for columns than for beams as the MRI changes from 700 years to 1700 years.

Peak inter-story drift and peak acceleration of the building were obtained from peak response databases (Figures 12 and 13) for MRI = 20 years and MRI = 10 years, respectively. Their maximum values in the two principal *x*- and *y*-axis and the associated resultant are summarized in Table 5. Note that inter-story drift and acceleration are not modified by the adjustment factor γ.

Table 4. Adjusted peak demand-to-capacity indexes

		MRI = 700 years		MRI = 1700 years	
		LC1	LC2	LC1	LC2
Corner	B_{ij}^{PM*}	0.94	0.74	1.04	1.01
column	B_{ij}^{VT*}	0.31	0.50	0.55	0.73
Non-corner	B_{ij}^{PM*}	1.00	0.74	1.08	0.84
column	B_{ij}^{VT*}	0.39	0.45	0.53	0.59
Exterior	B_{ij}^{PM*}	0.60	0.59	0.73	0.72
beam	B_{ij}^{VT*}	0.50	0.47	0.60	0.56
Interior	B_{ij}^{PM*}	0.67	0.66	0.79	0.78
beam	B_{ij}^{VT*}	0.64	0.60	0.75	0.70

5.7 Compliance with design criteria

Once peak structural responses for specified MRIs are obtained, DAD verifies if the peak responses satisfy design criteria for safety and serviceability. Figure 14 shows adjusted peak demand-to-capacity indexes accounting for ASCE 7 limitations on overturning moments. The indexes in the figure are the maxima of the load combination cases LC1 and LC2. They indicate that structural members were adequately designed for shear strength and have the capacity to resist effects of interacting shear forces and torsional moment (i.e., $B_{ij}^{VT^*} \leq 1$) corresponding to both MRIs. However, some members do not have adequate axial and flexural strengths (i.e., $B_{ij}^{PM} > 1$) for MRI = 1700 years. (A higher-than-unity index means than the corresponding member must be redesigned to achieve stronger capacity.) The overall DAD results show that structural members used in this study were designed more conservatively at higher floor levels, since the indexes typically decrease with height. Differences between peak responses corresponding to the two MRIs are member-dependent.

Table 5 shows peak inter-story drift ratios for MRI = 20 years and peak top floor accelerations for MRI = 10 years. The peak inter-story drift ratio based on DAD is 0.0029 in the y direction. The ASCE 7-05 Commentary suggests limits on the order of 1/600 to 1/400 (see Section CC.1.2 in ASCE 7-05). In this study this suggested criterion is not satisfied.

The peak top floor resultant acceleration based on DAD is 19.6 mg. This study assumed a limit of 25 mg for a 10-year MRI for office buildings (Isyumov et al. 1992). The limit is greater than the peak acceleration determined in this study. The design is therefore adequate for peak acceleration.

According to the strength criteria, the design is not adequate for the axial and flexural strength of columns for the 1700-year MRI. The design is also inadequate for peak inter-story drift corresponding to a 10-year MRI. Therefore, the procedure outlined in Sections 5.2 to 5.6 should be repeated with a modified structural design until the corresponding results satisfy the design criteria. This iterative, trial-and-error procedure is time-consuming. Therefore, an automated optimization procedure would be needed from the sake of both economy and computational efficiency. Such a procedure, which makes use of the DAD estimates of the response, is currently under development. Provision for P-delta effects is planned for a future version of the software.

Table 5. Peak inter-story drifts and peak acceleration

	ASCE 7		DAD		
	x dir.	y dir.	x dir.	y dir.	res.
Inter-story drift ratio [$\times 10^{-4}$]	13	23	17	29	31
Acceleration [mg]	12.7	19.5	15.0	16.4	19.6

Note: res. denotes resultant.

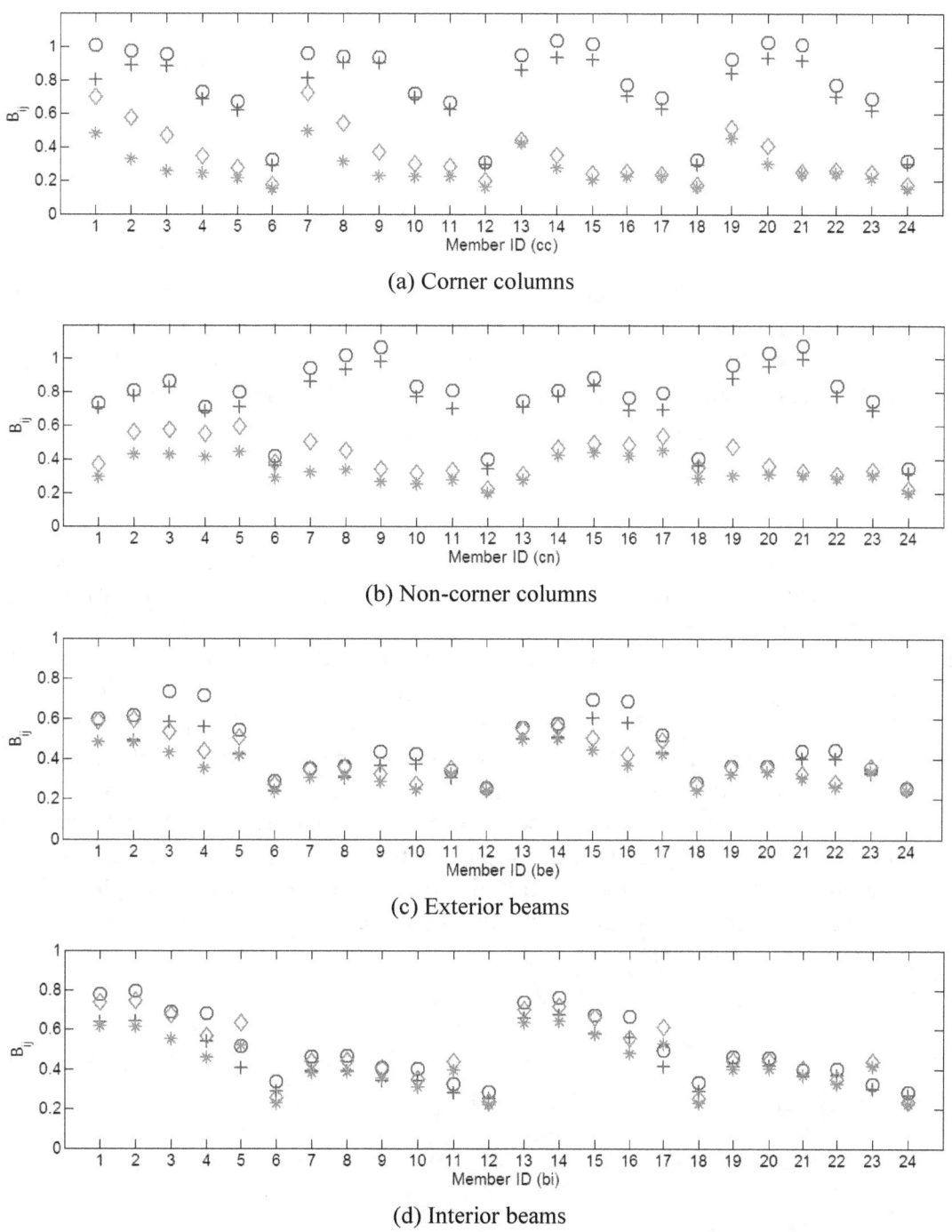

Figure 14. Design results for MRI = 700 and 1700 years (DAD)

(For MRI = 700 years, $\bigcirc B_{ij}^{PM*}$, $\diamondsuit B_{ij}^{VT*}$; for MRI = 1700 years, $+ B_{ij}^{PM*}$, $* B_{ij}^{VT*}$)

5.8 Comparisons of DAD- and ASCE 7-based designs

Figure 15 shows, for selected members, ratios of demand-to-capacity indexes based on DAD to those based on the ASCE 7 analytical method:

$$R = \frac{B_{ij}^{DAD} - B_{ij}^{ASCE7}}{B_{ij}^{ASCE7}} \tag{20}$$

where B_{ij}^{ASCE7} is the demand-to-capacity index from ASCE 7, and B_{ij}^{DAD} is the adjusted index from DAD.

The results indicate that differences between DAD- and ASCE 7-based results are significant in the column members. The ratio varies according to individual member, owing to the stronger dependence on individual members of the index based on DAD. For columns, the ASCE 7-based results overestimate B_{ij}^{PM}'s for lower floors, e.g., by approximately 30 % for corner columns in comparison with DAD-based results, but underestimate B_{ij}^{PM}'s for higher floors. ASCE 7 overestimates most B_{ij}^{VT} indexes, by up to approximately 65 % for lower floors. For beams, ASCE 7 overestimates the indexes at lower floors and underestimates them at higher floors, by up to approximately 20 %. These comparisons show that the ASCE 7 analytical method can result in structural members that are either stronger or weaker than the more realistically designed members based on DAD.

Maximum inter-story drift ratio for MRI = 20 years and acceleration for MRI = 10 years were also calculated by the ASCE 7-based method (Section C6.5.8 in ASCE 7-05), see Table 5, yielding a maximum inter-story drift ratio of 0.0023. This is less than the value obtained by DAD, which accounts for translational and rotational responses. The maximum acceleration is 19.5 mg in the y direction. This is close to the 19.6 mg resultant obtained by DAD. Note that the ASCE 7 method can yield peak inter-story drift or peak acceleration values lower than those yielded by DAD, meaning that the ASCE 7 estimates, based as they are on a physically simplified model, can be unconservative. The larger DAD drift and acceleration values can be explained in part by the fact that across-wind and torsional effects are accounted for in the DAD method but are disregarded by the ASCE 7 Standard.

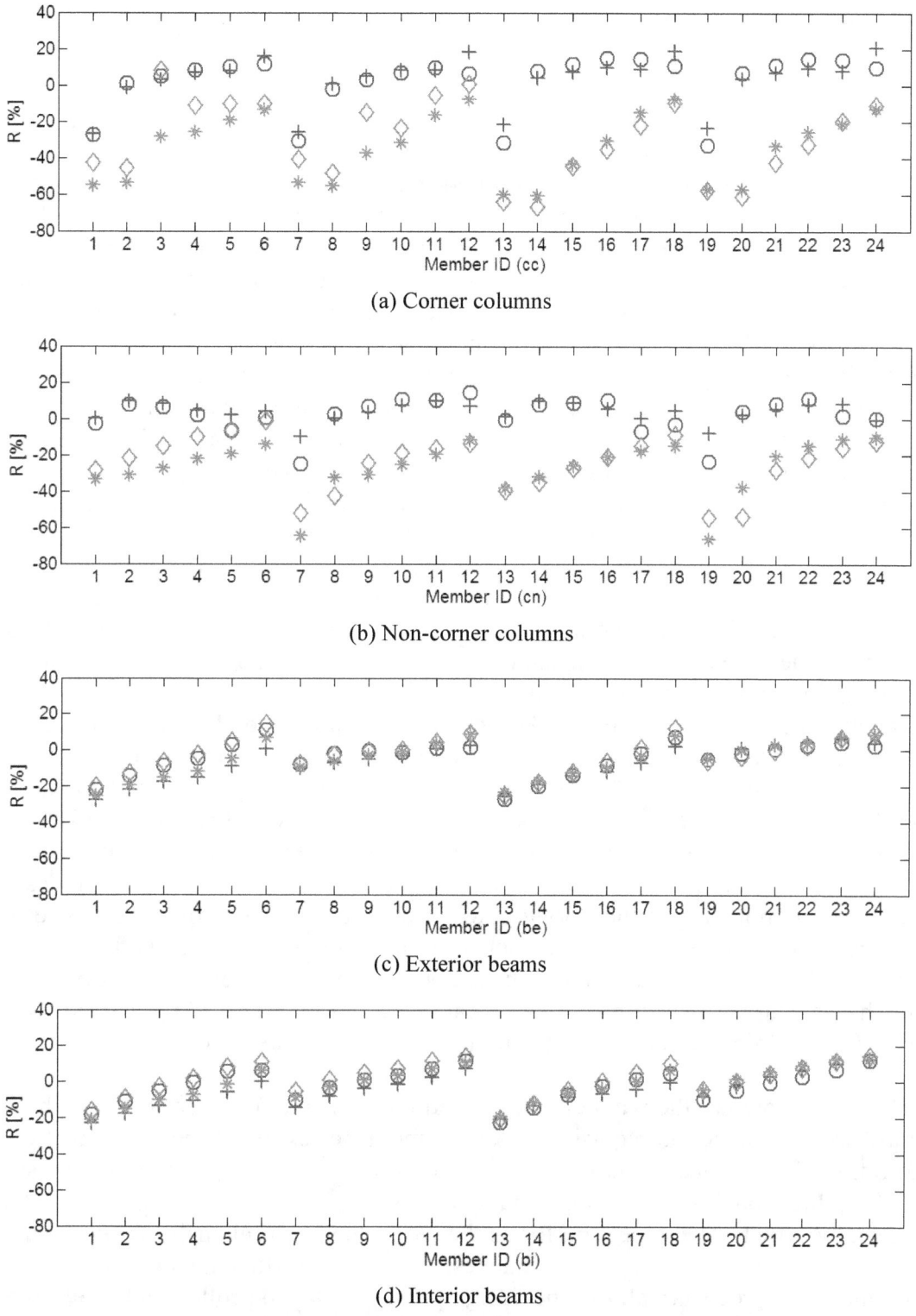

(a) Corner columns

(b) Non-corner columns

(c) Exterior beams

(d) Interior beams

Figure 15. Ratio R (Eq. (20))

(For MRI = 700 years, $\circ\ B_{ij}^{PM*}$, $\diamond\ B_{ij}^{VT*}$; for MRI = 1700 years, $+\ B_{ij}^{PM*}$, $*\ B_{ij}^{VT*}$)

6. Concluding remarks

This report presented the development of a Database-Assisted Design (DAD) procedure for reinforced concrete buildings, and its application to a 60-story building. The DAD procedure performs dynamic analyses using simultaneous time-series of aerodynamic pressure data obtained in the wind tunnel. It obtains displacement and acceleration time histories, effective lateral load time histories at the mass center at each floor, and time-series of demand-to-capacity indexes for axial force and moments, and shear force and torsion for any desired mean recurrence interval. Response databases for each index were established for a sufficiently wide range of wind speeds and for a sufficiently large number of wind directions. Response databases of inter-story drift and acceleration were also obtained. The databases depend on the building's aerodynamic, geometric, structural, and dynamical features but are independent of the wind climate.

The study employed directional wind speed data of hurricanes for a Miami location, obtained from the directional hurricane wind speed database listed on www.nist.gov/wind. Demand-to-capacity indexes were adjusted in accordance with ASCE 7-05 requirements.

The DAD methodology has the following advantages over frequency-domain procedures: (1) it preserves all phase relationships, so structural responses due to combined effects (e.g., combined effects in the directions of the principal axes of the building) are calculated by superposing individual effects; (2) wind loads along the building height are based on the actual distribution of the pressures as measured in the wind tunnel; (3) any modal shape, higher modes of vibration, and mode coupling are easily accounted for.

DAD appropriately accounts for wind directionality using wind climatological data that may need to be augmented by simulation, aerodynamic data, and micro-meteorological data (i.e., ratios of directional wind speeds in open exposure at 10 m above ground to their mean hourly counterparts at the top of the structure). Estimated peak responses obtained from DAD are estimated for the requisite mean recurrence intervals. This requires that the estimates be performed in the wind effects space. The procedure requires the use of structural engineering design information in the form, notably, of appropriate interaction equations specific to reinforced concrete members. Provision for P-delta effects is planned for a future version of the software.

The procedure was illustrated through its application to a specific design of the CAARC building. The conclusions based on this application would clearly differ for different types of building or design. Software for implementing the DAD procedure used in this study is available on www.nist.gov/wind.

DAD clearly separates the wind engineer's and the structural engineer's tasks. The wind engineer's task is to produce the requisite pressure time histories, wind climatological directional data, and ratios of directional wind speeds at standard elevation in open terrain exposure to the corresponding directional hourly mean wind speeds at the top of the structure. Once these data are available, the structural engineer performs the requisite structural analyses and accurately determines members' demand-to-capacity indexes, inter-story drift, and top floor accelerations. Therefore, the DAD procedure allows the design process to be controlled and scrutinized by the structural engineer. DAD renders the design process clear and transparent, and makes the partners in the design process clearly accountable to all stakeholders, including owners and building inspectors. The design approach presented in this paper provides more accurate and clearer predictions of wind effects than conventional approaches, and is expected to be more economical and efficient when used in conjunction with optimization.

References

ACI (2008). *Building code requirements for structural concrete (ACI 318-08) and commentary*, American Concrete Institute, Farmington Hills, MI.

ASCE (2005). *Minimum design loads for buildings and other structures*, American Society of Civil Engineers, Reston, VA.

Batts, M. E., Russell, L. R., and Simiu, E. (1980). "Hurricane wind speeds in the United States." *Journal of the Structural Division-ASCE*, 106(10), 2001-2016.

Coffman, B. F., Main, J. A., Duthinh, D., and Simiu, E. (2010). "Wind effects on low-rise buildings:Databased-Assisted Design vs. ASCE 7-05 Standard estimates." *Journal of Structural Engineering* (accepted).

Diana, G., Giappino, S., Resta, F., Tomasini, G., and Zasso, A. "Motion effects on the aerodynamic forces for an oscillating tower through wind tunnel tests." *5th European & African Conference on Wind Engineering*, Florence, Italy, 53-56.

Fritz, W. P., Bienkiewicz, B., Cui, B., Flamand, O., Ho, T. C. E., Kikitsu, H., Letchford, C. W., and Simiu, E. (2008). "International Comparison of Wind Tunnel Estimates of Wind Effects on Low-Rise Buildings: Test-Related Uncertainties." *Journal of Structural Engineering*, 134(12), 1887-1890.

Grigoriu, M. (2009). *Algorithms for generating large sets of synthetic directional wind speed data for hurricane, thunderstorm, and synoptic winds*. NIST Technical Note 1626, National Institute of Standards and Technology, Gaithersburg, MD.

Isyumov, N., Fediw, A. A., Colaco, J., and Banavalkar, P. V. (1992). "Performance of a tall building under wind action." *Journal of Wind Engineering and Industrial Aerodynamics*, 42(1-3), 1053-1064.

Kareem, A., Kijewski, T., and Tamura, Y. (1999). "Mitigation of motions of tall buildings with specific examples of recent applications." *Wind and Structures*, 2(3), 201-251.

Melbourne, W. H. (1980). "Comparison of measurements on the CAARC standard tall building model in simulated model wind flows." *Journal of Wind Engineering and Industrial Aerodynamics*, 6(1-2), 73-88.

NIST (2009). *High-Rise Database-Assisted Design*: www.nist.gov/wind.

PCA (2008). *PCA notes on 318-08 building code requirements for structural concrete with design applications*, Portland Cement Association, Skokie, IL.

Simiu, E., Gabbai, R. D., and Fritz, W. P. (2008). "Wind-induced tall building response: a time-domain approach." *Wind and Structures*, 11(6), 427-440.

Simiu, E., and Miyata, T. (2006). *Design of buildings and bridges for wind: a practical guide for ASCE-7 Standard users and designers of special structures*, John Wiley & Sons, Hoboken, NJ.

SOM (2004). "WTC wind load estimates, outside experts for baseline structural performance Appendix D." in *NIST NCSTAR1-2, Baseline structural performace and aircraft impact damage analysis of the World Trade Center towers*, submitted by Skidmore, Owings and Merrill LLP, Chicago, Illinois, 13 April 2004 (wtc.nist.gov), also reproduced as Appendix to NIST Technical Note 1655, "Toward a standard on the wind tunnel method" (2009) by E. Simiu, pp. A1-A7 (www.nist.gov/wind).

Spence, S. M. J. (2009). *High-rise database-assisted design 1.1 (HR_DAD_1.1): Concepts, software, and examples*. NIST Building Science Series 181, National Institute of Standards and Technology, Gaithersburg, MD.

Teshigawara, M. (2001). "Structural design principles (chapter 6)." in *Design of modern highrise reinforced concrete structures*, H. Aoyama, ed., Imperial College Press, London.

Venanzi, I. (2005). *Analysis of the torsional response of wind-excited high-rise building*, Ph.D. Dissertation, Università degli Studi di Perugia.

Wardlaw, R. L., and Moss, G. F. "A standard tall building model for the comparison of simulated natural winds in wind tunnels." *International conference on wind effects on buildings and structures*, Tokyo, Japan, 1245-1250.

Yeo, D., and Simiu, E. (2010). *Effects of veering wind and structure orientation on a high-rise structure*. NIST Technical Note National Institute of Standards and Technology, Gaithersburg, MD. (in preparation).

Appendix:

User's Manual

HR_DAD_RC, version 1.0
High-Rise Database-Assisted Design
for Reinforced Concrete Structures

Appendix

High-Rise Database-Assisted Design Software
for Reinforced Concrete Structures
(HR_DAD_RC, version 1.0)

User's Manual

developed by DongHun Yeo

Building and Fire Research Laboratory

National Institute of Standards and Technology

Updated May 1, 2010

Current version available at www.nist.gov/wind

Disclaimer

Certain trade names or company products are specified in this document to specify adequately the procedure sued. Such identification does not imply recommendation or endorsement by NIST, nor does it imply that the product is the best available for the purpose. The "stand-alone" version of this software requires installation of the MATLAB [1] Component Runtime (MCR) Libraries provided by The MathWorks, Inc. The author's limited rights to the deployment of this program are limited by a license agreement between NIST and The MathWorks. The license agreement can be found at www.mathworks.com/license/. The author, NIST, and The MathWorks and its licensors are excluded from all liability for damages or any obligation to provide remedial actions.

---

[1] MATLAB®. © 1984 - 2009 The MathWorks, Inc.

1. New features of HR_DAD_RC

This software is the first version of HR_DAD for reinforced concrete structures. Previous HR_DAD programs have been developed for steel structures. HR_DAD_RC adds programs for reinforced concrete structures and significantly improves the performance of the entire HR_DAD software.

1.1 General features

Like previous versions of HR_DAD, HR_DAD_RC implements a procedure and format which accounts for wind directionality and calculates demand-to-capacity indexes corresponding to the same specified mean recurrence interval for all individual member of concern (for details see Simiu et al. (2008)). The capability is unique to HR_DAD.

- *Estimation of peaks of member demand-to-capacity indexes*
  A demand-to-capacity index is the sum of ratios of internal forces (e.g., moment and axial force) divided by the respective capacities. (This sum constitutes the left-hand side of the interaction equation used in the design of individual members.) Two options are available. The default option sums up the ratios for each time step of the full time series of the internal forces based on observed peaks. A much faster but accurate alternative option uses an updated point-in-time approach where the ratios are obtained for a limited number of points in time (e.g., 5 to 100).

- *Number of internal forces in expression for member demand-to-capacity index*
  Instead of just three internal forces (axial force, moments about the principal axes of the member) as in the HR_DAD software for steel structures, six internal forces are considered (axial force, shear forces for two axes, torsional moment, and bending moments for two axes).

- *Number of demand-to-capacity indexes*
  In addition to the demand-to-capacity index for axial force and bending moments, an index for shear forces and torsion is calculated for structural members.

- *Effects of veering angle*
  Wind directions at top of a high-rise building are different from those at 10 m above ground due to veering (change of wind direction with elevation). HR_DAD_RC has the capability for accounting for the effect of veering on the demand-to-capacity indexes.

- *Directional terrain exposure*
  Actual terrain exposures at the location of concern as functions of azimuth angles are taken into account when converting reference speeds measured at meteorological stations to mean wind speeds at the elevation of the top of the building. The converted speeds are used in conjunction with aerodynamic databases obtained by accounting for the direction-dependent terrain exposure.

- *Building orientation*
  The orientation angle of a structure is accounted for in the calculation of the response. If the option of choosing the building orientation is available, the designer can determine the orientation

of the structure that is optimal from a structural performance point of view. For this option to be used directional aerodynamic data must be available for each individual building orientation.

1.2 Special features for RC

- *Section details of RC members*
  RC members consist of beams, columns, and shear walls. Each type of member has different features (e.g., width, depth, reinforcement details). HR_DAD_RC covers beams with either tension reinforcement only or tension and compression reinforcement. Columns are based on symmetric distribution of reinforcement in a rectangular section. Shear wall modeling is not covered in the present report and is the object of an ongoing study.

- *Demand-to-capacity indexes*
  These indexes are specific to RC structures.

2. Download and installation

HR_DAD_RC can be accessed via the internet site http://www.nist.gov/wind. Under the heading "II. Wind Effects on Buildings," click the link "Wind Effects on Flexible Buildings". This opens the main page "HR_DAD – DAD Software for High-Rise Buildings". The files are available in a bulleted list under the heading *"Files available for download"*. In the following, the name of the associated bullet is used for each set to files being downloaded.

From that list, consider first the bullet "Files for HR_DAD software." Next to the title of the bullet item is a link to the self-extracting file zip file "hr_dad_for_RC.exe", which contains the 18 MATLAB files required to run the software. The user should download this *.exe file[2] and proceed to extract the 18 files to a folder that the user specifies to store the files (e.g., C:\HR_DAD_RC). The user can add this directory to the MATLAB search path if necessary.

The user can run HR_DAD_RC by typing 'HR_DAD' in the MATLAB command window when the cursor is in the designated folder where the extracted files are located. The program opens a window consisting of three pages ('Modeling', 'Wind Effects', and 'Results & Plots') that form the graphical user interface (GUI).

3. User's Guide

HR_DAD_RC has three main pages (Modeling, Wind Effects, and Results & Plot). The 'Modeling' page is used to assign values to the variables used to model building and loads. The 'Wind Effect' page is used

---

[2] Users using UNIX/Linux platforms should download the non-self-extracting.zip file "hr_dad_rc.zip".

to perform the requisite response databases and the associated peak responses with specified MRIs (Mean Recurrence Intervals). The 'Results & Plots' page is used to show the results and the associated graphs.

Variable values can be loaded from a pre-assigned file by clicking 'Load' button at the left bottom corner in any page. They also can be saved into a file by executing 'Save' button at the right bottom corner in any page. The button of 'Exit' next to the 'Save' one enables to stop the program by clicking it.

Main computations of HR_DAD_RC software are performed by two separate script files, "Program1.m" and "Program2.m", which are executed consecutively. Once all the variables are assigned in 'Modeling', the first script file of the HR_DAD_RC, "Program1.m" performs dynamic analysis of a building and calculates its response database by clicking the "Compute response database" button. This process requires (1) the aerodynamic database of pressure time-histories under wind in each direction obtained from wind tunnel testing or CFD procedures, and (2) relevant influence coefficients that can be obtained from any commercial or in-house FEA programs. The results of this first computation are databases of peak wind effects in each wind direction (e.g., demand-to-capacity index for strength design, inter-story drift, and acceleration) for a sufficient number of wind speeds. They are named and saved in the locations specified at the top of the 'Wind Effects' page.

The second script file "Program2.m" calculates, for specified MRIs, peak directional wind effects of the building at a given geographical location. To account for wind climatological data at the building site, a climatological database of hurricane and/or non-hurricane wind speeds for the location of interest is needed. A set of simulated hurricane wind speeds may be downloaded via the link "Simulated Directional Hurricane Wind Speed Data" bullet. The wind dataset must be saved in a folder, which is selected in the 'Peak Responses' section of the 'Wind Effects' page. Once the directory containing the simulated hurricane wind speed files is selected, the appropriate milepost can be chosen from the climatological database. The chosen milepost must correspond to the wind speeds at or near the building location. Other directional hurricane wind speed databases may be used if available. For non-hurricane wind climates observed wind speed data sets are used, which are augmented by numerical simulation as shown, e.g., in Grigoriu (2009). Also required are micro-meteorological data obtained in the wind tunnel, consisting of the ratio between the directional mean hourly wind speeds at the elevation of the top of the building and the corresponding directional 3-s or 10-min wind speeds at the standard elevation (typically 10 m) of the meteorological site.

Once both wind climatological and micro-meteorological data are available and "Program1.m" is executed, "Program2.m" can be executed by clicking "Compute peak responses with specified MRIs" on the bottom of the 'Wind Effects' page. The results of this second run are the peak directional wind effects for a specified MRI that account for wind climatological data, building orientation, and directional surface roughness around the building. They are named and saved in the locations specified on the bottom of the 'Wind Effects' page. Structural engineers must provide dynamic characteristics and influence coefficients of the building. Wind engineers must provide the appropriate aerodynamic database, wind climatological database, and micro-meteorological data set.

Finally, the page 'Results & Plots' allows the user to view the results of response database and peak structural responses with specified MRIs obtained by executing "Program1.m" and "Program2.m".

■ Example of a 3D 60-story RC building

Consider a 60-story building (i.e., the CAARC building (Melbourne 1980)), with floors assumed to be rigid diaphragms (Figure A-1). The building is 45.72 m (150 ft) in width (B), 30.48 m (100 ft) in depth (D), and 182.88 m (600 ft) in height (H). It has a moment-resistant frame structure that consists of 2880 columns and 4920 beams.

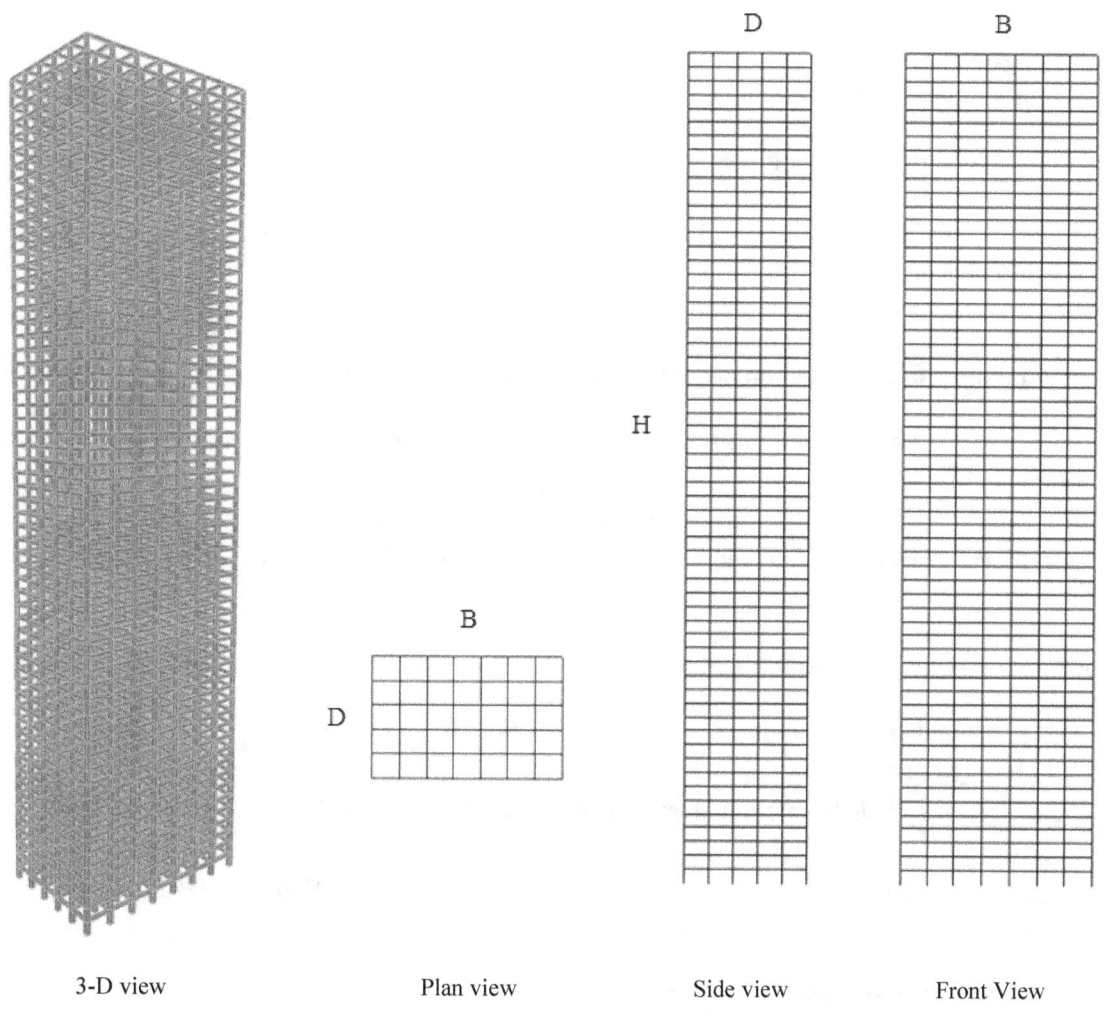

3-D view   Plan view   Side view   Front View

Figure A-1. Schematic views of a 60-story building

1) Input for Running HR_DAD for RC

HR_DAD requires as input information including details on the structural members, wind-induced pressure coefficients on the external surfaces, and wind climatological data at the location of the structure, as follows:

- Information related to details on structural members:

  - Dimensions of members and their reinforcement details

- Information related to the dynamic properties of the structure:

  - Mass and mass moment of inertia with respect to mass center for each floor
  - Mode shapes, frequencies, and damping ratios

- Information related to structural loadings:

  - Time varying loads applied at mass center of each floor, obtained from wind tunnel tests using a rigid model
  - Static loads applied to each member

- Information related to the wind climate at or near the building site:

  - Databases of directional wind speeds for each type of wind occurring at the site (e.g., hurricane, thunderstorm, synoptic winds)
  - Information regarding the ratio between 3s-averaged wind speeds at 10m above in open terrain and their 1hr mean wind speed counterparts at the top of the building for the terrain exposure at the site.

2) Output from running HR_DAD_RC

  - Peak demand/capacity indexes for all members, corresponding to specified Mean Recurrence Intervals (MRIs)
  - Peak inter-story drift calculated at any point within column lines, corresponding to specified MRIs
  - Peak top floor acceleration calculated for corners of the top floor, corresponding to specified MRIs

3) Details on Use of the Software

The following pages illustrate in detail how to input the information required for HR_DAD_RC. General description of input variables and their associated format is provided in the subsequent pages.

Building Modeling:

```
────────────────────────────── Building Modeling ──────────────────────────────

Basic infomation        Building height [m]:        No. of stories:           No. of DOFs per floor:
                        180.22                      60                        3

Structural properties   List of all members:                                  Details of all members:
                        C:\HR_DAD_RC\Input\RC_members_list.mat   [Select file...]   C:\HR_DAD_RC\Input\RC_member_properties.mat   [Select file...]

                        Heights of floors:
                        C:\HR_DAD_RC\Input\height_floors.mat   [Select file...]

                        Influence coefficients:
                        C:\HR_DAD_RC\Input\iif_all.mat   [Select file...]

Dynamic properties      No. of modes:               Modal periods [s]:        Mode shapes:
                        3                           6.05   5.7   5             C:\HR_DAD_RC\Input\ModeShapes.mat   [Select file...]

                        Modal damping ratio [%]:                               Mass:
                        2 2 2                                                  C:\HR_DAD_RC\Input\mass_asc.mat   [Select file...]
```

Basic information	
Building height	= Height of the building being designed (m) → 182.88 (m) in the example
No. of stories	= Number of stories → 60 (stories)
No. of DOFs per floor	= Degrees of freedom per floor → 3 (DOFs)
Structural properties	
List of members	= Load the MATLAB file containing list of members of the structure The variable is named "**mem_list**" and can be saved in a mat file with an arbitrary name. The variable is a matrix with three rows and number of columns that make up the structure. The first row contains the member numbers. The second row has types of members using string characters (i.e., "C", "B", or "W" that represents column, beam, and shear wall, respectively.) The third row contains the identifier of the member defined in the second row (e.g., 1 for "C" means column1, and 2 for "B" means beam2.) **mem_list** (3, No. of members) = $$\begin{bmatrix} 1 & 2 & \dots & n & \dots & \text{No. of members} \\ \text{"C"} & \text{"C"} & \dots & \text{"B"} & \dots & \text{"B"} \\ 1 & 1 & \dots & 2 & \dots & 10 \end{bmatrix}$$ → **mem_list** (3, 7800) is saved in RC_members_list.mat. The variable contains

	12 types of columns and beams, respectively.
Details of members	= Load the MATLAB file containing the section properties The file must have the following properties of the reinforced concrete sections: • *Width of members (b_member):* This variable is a matrix for member width that consists of rows of three member types, and of columns whose number is the largest number of identifiers for any of the member types. The identifier of the members is defined in the list of members. **b_member** (3, max. no. of identifiers in any type) = $$\begin{array}{c} \text{columns} \\ \text{beams} \\ \text{shear walls} \end{array} \begin{bmatrix} b_{C1} & b_{C2} & .... & b_{Cm} \\ b_{B1} & b_{B2} & .... & b_{Bn} \\ b_{W1} & b_{W2} & .... & b_{Wo} \end{bmatrix}$$ → **b_member** (3, 12) is saved in RC_members_properties.mat. The unit is mm. Note: If the number of identifiers is different for the various types of members (columns, beams, and shear walls), the largest value of any of the types of members decides the column dimension of the variable. For member types with fewer identifiers, the value beyond their identifier number is defined as zero. • *Height of members (h_member):* This variable is a matrix for member height that consists of rows of three member types, and of columns whose number is the largest number of identifiers for any of the member types. The identifier of the members is defined in the list of members. **h_member** (3, max. no. of identifiers in any type) = $$\begin{array}{c} \text{columns} \\ \text{beams} \\ \text{shear walls} \end{array} \begin{bmatrix} h_{C1} & h_{C2} & .... & h_{Cm} \\ h_{B1} & h_{B2} & .... & h_{Bn} \\ h_{W1} & h_{W2} & .... & h_{Wo} \end{bmatrix}$$ → **h_member** (3, 12) is saved in RC_members_properties.mat. The unit is mm. Note: If number of identifiers is different for various types of members (columns, beams, and shear walls), the largest value of any of the types of member decides the column dimension of the variable. For member types with fewer identifiers, the value beyond their identifier number is defined as zero. • *Compression strength of concrete (fc_conc):* The variable is a matrix for compression strength of concrete. It consists of rows of three member types and columns whose number is the largest number of identifiers for any of the member types. The identifier of members is defined in

the list of members.

fc_conc (3, max. no. of identifiers in any type) =

$$
\begin{array}{c}
\text{columns} \\
\text{beams} \\
\text{shear walls}
\end{array}
\begin{bmatrix}
f_{c\_C1} & f_{c\_C2} & \cdots & f_{c\_Cm} \\
f_{c\_B1} & f_{c\_B2} & \cdots & f_{c\_Bn} \\
f_{c\_W1} & f_{c\_W2} & \cdots & f_{c\_Wo}
\end{bmatrix}
$$

→ **fc_conc** (3, 12) is saved in RC_members_properties.mat. The unit is MPa.

Note: For member types with fewer identifiers, the value beyond their identifier number is defined as zero.

- *Lightweight aggregate concrete factor (lambda):*
The variable is a matrix for the factor of lightweight aggregate concrete defined in Section 5.8, ACI 318-08. It consists of rows of three member types and columns whose number is the largest number of identifiers for any of the member types. The identifier of the members is defined in the list of members.

lambda (3, max. no. of identifiers in any type) =

$$
\begin{array}{c}
\text{columns} \\
\text{beams} \\
\text{shear walls}
\end{array}
\begin{bmatrix}
\lambda_{C1} & \lambda_{C2} & \cdots & \lambda_{Cm} \\
\lambda_{B1} & \lambda_{B2} & \cdots & \lambda_{Bn} \\
\lambda_{W1} & \lambda_{W2} & \cdots & \lambda_{Wo}
\end{bmatrix}
$$

→ **lambda** (3, 12) is saved in RC_members_properties.mat. The value of unity is used for normal weight concrete.

Note: For member types with fewer identifiers, the value beyond their identifier number is defined as zero.

- *Yield strength of longitudinal reinforcement (fy_st):*
The variable is a matrix for yield strength of reinforcement. It consists of rows of three member types and columns whose number is the largest number of identifiers for any of the member types. The identifier of the members is defined in the list of members.

fy_st (3, max. no. of identifiers in any of the types of member) =

$$
\begin{array}{c}
\text{columns} \\
\text{beams} \\
\text{shear walls}
\end{array}
\begin{bmatrix}
f_{st\_C1} & f_{st\_C2} & \cdots & f_{st\_Cm} \\
f_{st\_B1} & f_{st\_B2} & \cdots & f_{st\_Bn} \\
f_{st\_W1} & f_{st\_W2} & \cdots & f_{st\_Wo}
\end{bmatrix}
$$

→ **fy_st** (3, 12) is saved in RC_members_properties.mat. The unit is MPa.

Note: For member types with fewer identifiers the value beyond their identifier number is defined as zero.

A10

● *Area of tension reinforcement (As1):*
The variable is a matrix for area of tension reinforcement. It consists of rows of three member types and columns whose number is the largest number of identifiers for any of the member types. The identifier of the members is defined in the list of members.

As1 (3, max. no. of identifiers in any type) =

$$
\begin{array}{l}
\text{columns} \\
\text{beams} \\
\text{shear walls}
\end{array}
\left[
\begin{array}{cccc}
As_{1\_C1} & As_{1\_C2} \dots & , & As_{1\_Cm} \\
As_{1\_B1} & As_{1\_B2} \dots & & As_{1\_Bn} \\
As_{1\_W1} & As_{1\_W2} \dots & & As_{1\_Wo}
\end{array}
\right]
$$

→ **As1** (3, 12) is saved in RC_members_properties.mat. The unit is mm^2.

Note 1: In this manual "tension reinforcement" and "compression reinforcement" for columns are described below. The tension and compression reinforcement are defined as reinforcement closest to the tension face and the compression face, respectively. The reinforcement can consist of more than one layer.

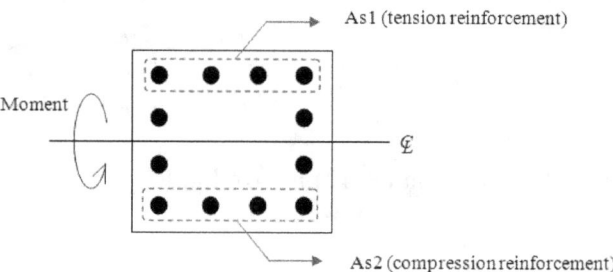

Note 2: For member types with fewer identifiers the value beyond their identifier number is defined as zero.

Note 3: Only a single layer of tension reinforcement is taken into account in this program. When double layers of tension reinforcement are used, the area must be calculated as for a single layer.

● *Area of compression reinforcement (As2):*
The variable is a matrix for area of compression reinforcement. It consists of rows of three member types and columns whose number is the largest number of identifiers for any of the member types. The identifier of the members is defined in the list of members.

As2 (3, max. no. of identifiers in any type) =

$$
\begin{array}{l}
\text{columns} \\
\text{beams} \\
\text{shear walls}
\end{array}
\left[
\begin{array}{ccc}
As_{2\_C1} & As_{2\_C2} \dots & As_{2\_Cm} \\
As_{2\_B1} & As_{2\_B2} \dots & As_{2\_Bn} \\
As_{2\_W1} & As_{2\_W2} \dots & As_{2\_Wo}
\end{array}
\right]
$$

→ **As2** (3, 12) is saved in RC_members_properties.mat. The unit is mm^2.

Note 1: For member types with fewer identifiers, the value beyond their identifier number is defined as zero.

Note 2: Only a single layer of compression reinforcement is taken into account in this program. When double layers of tension reinforcement are used, the area must be calculated as for a single layer.

● *Area of total longitudinal reinforcement (As_ttl):*
The variable is a matrix for area of total longitudinal reinforcement. It consists of rows of three member types and columns whose number is the largest number of identifiers for any of the member types. The identifier of the members is defined in the list of members.

As_ttl (3, max. no. of identifiers in any types) =

$$\begin{array}{l} \text{columns} \\ \text{beams} \\ \text{shear walls} \end{array} \begin{bmatrix} As_{t\_C1} & As_{t\_C2} & \dots & As_{t\_Cm} \\ As_{t\_B1} & As_{t\_B2} & \dots & As_{t\_Bn} \\ As_{t\_W1} & As_{t\_W2} & \dots & As_{t\_Wo} \end{bmatrix}$$

→ **As_ttl** (3, 12) is saved in RC_members_properties.mat. The unit is mm^2.

Note 1: For member types with fewer identifiers, the value beyond their identifier number is defined as zero.

Note 2: Area of total longitudinal reinforcement is not always the sum of the area of all reinforcing and compression reinforcement. When the section of a column has intermediate reinforcement between tension and compression reinforcement, the total area must include the area of all longitudinal reinforcement such as tension, compression, and intermediate reinforcements.

● *Distance from end part of concrete to center of tension reinforcement (d_1):*
The variable is a matrix for the distance from the end part of concrete to center of tension reinforcement. The distance is described in ACI 318-08. It consists of (1) rows of three member types and (2) columns whose number is the largest number of the identifiers in any member types. The identifier of the members is defined in the list of members.

d_1 (3, max. no. of identifiers in any type) =

$$\begin{array}{l} \text{columns} \\ \text{beams} \\ \text{shear walls} \end{array} \begin{bmatrix} d_{1\_C1} & d_{1\_C2} & \dots & d_{1\_Cm} \\ d_{1\_B1} & d_{1\_B2} & \dots & d_{1\_Bn} \\ d_{1\_W1} & d_{1\_W2} & \dots & d_{1\_Wo} \end{bmatrix}$$

→ **d_1** (3, 12) is saved in RC_members_properties.mat. The unit is mm.

Note: For member types with fewer identifiers, the value beyond their identifier number is defined as zero.

- *Distance from end part of concrete to center of compression reinforcement (d_2):*

The variable is a matrix for the distance from the end part of concrete to the center of compression reinforcement. The distance is described in ACI 318-08. It consists of rows of three member types and columns whose number is the largest number of identifiers for any of the member types. The identifier of the members is defined in the list of members.

d_2 (3, max. no. of identifiers in any type) =

$$
\begin{array}{c}
\text{columns} \\
\text{beams} \\
\text{shear walls}
\end{array}
\left[
\begin{array}{cccc}
d_{2\_C1} & d_{2\_C2} & \cdots & d_{2\_Cm} \\
d_{2\_B1} & d_{2\_B2} & \cdots & d_{2\_Bn} \\
d_{2\_W1} & d_{2\_W2} & \cdots & d_{2\_Wo}
\end{array}
\right]
$$

→ **d_2** (3, 12) is saved in RC_members_properties.mat. The unit is mm.

Note: For member types with fewer identifiers, the value beyond their identifier number is defined as zero.

- *Distance from end part of concrete to center of extreme tension reinforcement (d_b):*

The variable is a matrix for the distance from the end part of concrete to center of extreme tension reinforcement. The distance is described in ACI318-08. It consists of rows of three member types and columns whose number is the largest number of identifiers for any of the member types. The identifier of the members is defined in the list of members.

d_b (3, max. no. of identifiers in any type) =

$$
\begin{array}{c}
\text{columns} \\
\text{beams} \\
\text{shear walls}
\end{array}
\left[
\begin{array}{cccc}
0 & 0 & \cdots & 0 \\
d_{b\_B1} & d_{b\_B2} & \cdots & d_{b\_Bn} \\
0 & 0 & \cdots & 0
\end{array}
\right]
$$

→ **d_b** (3, 12) is saved in RC_members_properties.mat. The unit is mm.

Note 1: For member types with fewer identifiers, the value beyond their identifier number is defined as zero.

Note 2: The variable is only for beams. When double-layer tension reinforcement is used, variables of d_b and d_1 are not identical.

- *Yield strength of shear reinforcement (fy_st_v):*
The variable is a matrix for yield strength of shear reinforcement. It consists of

(1) rows of three member types and (2) columns whose number is the largest number of the identifiers for any of the member types. The identifier of the members is defined in the list of members.

fy_st_v (3, max. no. of identifiers in any types) =

$$
\begin{array}{l}
\text{columns} \\
\text{beams} \\
\text{shear walls}
\end{array}
\left[
\begin{array}{cccc}
f_{st\_v\_C1} & f_{st\_v\_C2} & \cdots & f_{st\_v\_Cm} \\
f_{st\_v\_B1} & f_{st\_v\_B2} & \cdots & f_{st\_v\_Bn} \\
f_{st\_v\_W1} & f_{st\_v\_W2} & \cdots & f_{st\_v\_Wo}
\end{array}
\right]
$$

→ **fy_st_v** (3,12) is saved in RC_members_properties.mat. The unit is MPa.

Note: For member types with fewer identifiers the value beyond their identifier number is defined as zero.

● *Area of shear reinforcement (Av):*
The variable is a matrix for area of shear reinforcement. It consists of rows of three member types and columns whose number is the largest number of the identifiers for any of the member types. The identifier of the members is defined in the list of members.

Av (3, max. no. of identifiers in any types) =

$$
\begin{array}{l}
\text{columns} \\
\text{beams} \\
\text{shear walls}
\end{array}
\left[
\begin{array}{cccc}
Av_{C1} & Av_{C2} & \cdots & Av_{Cm} \\
Av_{B1} & Av_{B2} & \cdots & Av_{Bn} \\
Av_{W1} & Av_{W2} & \cdots & Av_{Wo}
\end{array}
\right]
$$

→ **Av** (3,12) is saved in RC_members_properties.mat. The unit is mm^2.

Note: For member types with fewer identifiers the value beyond their identifier number is defined as zero.

● *Spacing of shear reinforcement (s):*
The variable is a matrix for spacing of shear reinforcement. It consists of rows of three member types and columns whose number is the largest number of identifiers for any of the member types. The identifier of the members is defined in the list of members.

s_v (3, max. no. of identifiers in any types) =

$$
\begin{array}{l}
\text{columns} \\
\text{beams} \\
\text{shear walls}
\end{array}
\left[
\begin{array}{cccc}
s_{v\_C1} & s_{v\_C2} & \cdots & s_{v\_Cm} \\
s_{v\_B1} & s_{v\_B2} & \cdots & s_{v\_Bn} \\
s_{v\_W1} & s_{v\_W2} & \cdots & s_{v\_Wo}
\end{array}
\right]
$$

→ **s_v** (3,12) is saved in RC_members_properties.mat. The unit is mm.

Note: For member types with fewer identifiers the value beyond their identifier number is defined as zero.

● *Area enclosed by the centerline of the outmost closed stirrups (Aoh):*
The variable is a matrix for area enclosed by the centerline of the outmost closed stirrups defined in ACI 318-08 (2008). It consists of rows of three member types and columns whose number is the largest number of identifiers for any of the member types. The identifier of the members is defined in the list of members.

Aoh (3, max. no. of identifiers in any types) =

$$
\begin{matrix}
\text{columns} \\
\text{beams} \\
\text{shear walls}
\end{matrix}
\begin{bmatrix}
A_{oh\_C1} & A_{oh\_C2} & \cdots & A_{oh\_Cm} \\
A_{oh\_B1} & A_{oh\_B2} & \cdots & A_{oh\_Bn} \\
A_{oh\_W1} & A_{oh\_W2} & \cdots & A_{oh\_Wo}
\end{bmatrix}
$$

→ **Aoh** (3,12) is saved in RC_members_properties.mat. The unit is mm.

Note: For member types with fewer identifiers the value beyond their identifier number is defined as zero.

● *Perimeter enclosed by the centerline of the outmost closed stirrups (Ph):*
The variable is a matrix for perimeter enclosed by the centerline of the outmost closed stirrups defined in ACI 318-08. It consists of rows of three member types and columns whose number is the largest number of identifiers for any of the member types. The identifier of the members is defined in the list of members.

Ph (3, max. no. of identifiers in any types) =

$$
\begin{matrix}
\text{columns} \\
\text{beams} \\
\text{shear walls}
\end{matrix}
\begin{bmatrix}
P_{h\_C1} & P_{h\_C2} & \cdots & P_{h\_Cm} \\
P_{h\_B1} & P_{h\_B2} & \cdots & P_{h\_Bn} \\
P_{h\_W1} & P_{h\_W2} & \cdots & P_{h\_Wo}
\end{bmatrix}
$$

→ **Ph** (3,12) is saved in RC_members_properties.mat. The unit is mm.

Note: For member types with fewer identifiers the value beyond their identifier number is defined as zero.

Heights of floors	= Load the MATLAB file containing the heights of all floors The variable is named "**H_floor**" and can be saved in a mat file with an arbitrary name. The variable is a row vector containing the heights of all floors and roof except for the ground floor. For example, the heights of floors in a 60-story building is: [3.048, 6.096, … , 182.88]

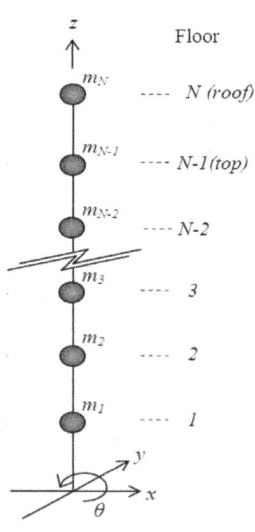

H_floor (No. of stories, 1) =
      [height at 1st floor, height at 2nd floor, ... , height at roof]

→ The variable **H_floor** (1,60) = [3.084 : 3.048 : 182.88] is saved in height_floors.mat. The unit is m.

Note: The figure above shows an N-story building that has mass (m_1, m_2, ... , m_N) at each floor. The heights of the floors are along the z axis in the figure above.

Influence coefficients	= Load the MATLAB file containing the influence coefficients The variable name must be "**dif**" and can be saved in a mat file with an arbitrary name. The variable is a 3D array. The following description of **dif** is in reference to the member showed in the figure below 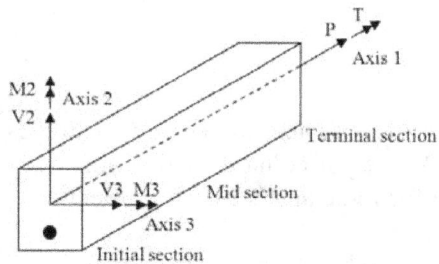 Each face of **dif** contains the influence coefficients associated with the six internal forces and moments (axial forces P, shear forces V2 and V3, torsion T, and moments M2 and M3) that are induced by a unit force or moment applied to the mass center of a given floor in one of the directions x, y, or θ. The internal forces are assigned from the first column to the sixth column such as P, V2, V3, T, M2, and M3. The first 3×Nfloors rows (Nfloors is the number of floors of the structure) contains influence coefficients in the initial section. The second and the third 3×Nfloors rows are for the influence coefficients in the mid section and in

the terminal section, respectively. Each Nfloors row contains influence coefficients in the x direction first from the first story to the highest story, then those in the y direction, and finally those in the θ direction. This makes up a total of 3×Nfloors rows for the influence coefficients of the each section. The total 9×Nfloors rows are assigned in every column (i.e., force or moment) for a given face. The index of each face identifies the member. For example, the ith face must correspond to the member described in the ith column of the variable **mem_list**. The array size is (9×Nfloors, 6, No. of members).

In an ith member, mem_list (: , : , i) =

	P	V2	V3	T	M2	M3
Initial section	I_{P_x1}	I_{V2_x1}	I_{V3_x1}	I_{T_x1}	I_{M2_x1}	I_{M3_x1}
	I_{P_x2}	I_{V2_x1}	I_{V3_x1}	I_{T_x1}	I_{M2_x1}	I_{M3_x1}
	$\vdots$	$\vdots$	$\vdots$	$\vdots$	$\vdots$	$\vdots$
	$I_{P\_xNfloors}$	$I_{V2\_xNfloors}$	$I_{V3\_xNfloors}$	$I_{T\_xNfloors}$	$I_{M2\_xNfloors}$	$I_{M3\_xNfloors}$
3 Nfloors	I_{P_y1}	I_{V2_y1}	I_{V3_y1}	I_{T_y1}	I_{M2_y1}	I_{M3_y1}
	I_{P_y2}	I_{V2_y1}	I_{V3_y1}	I_{T_y1}	I_{M2_y1}	I_{M3_y1}
	$\vdots$	$\vdots$	$\vdots$	$\vdots$	$\vdots$	$\vdots$
	$I_{P\_yNfloors}$	$I_{V2\_yNfloors}$	$I_{V3\_yNfloors}$	$I_{T\_yNfloors}$	$I_{M2\_yNfloors}$	$I_{M3\_yNfloors}$
	$I_{P\_\theta1}$	$I_{V2\_\theta1}$	$I_{V3\_\theta1}$	$I_{T\_\theta1}$	$I_{M2\_\theta1}$	$I_{M3\_\theta1}$
	$I_{P\_\theta2}$	$I_{V2\_\theta1}$	$I_{V3\_\theta1}$	$I_{T\_\theta1}$	$I_{M2\_\theta1}$	$I_{M3\_\theta1}$
	$\vdots$	$\vdots$	$\vdots$	$\vdots$	$\vdots$	$\vdots$
	$I_{P\_\theta Nfloors}$	$I_{V2\_\theta Nfloors}$	$I_{V3\_\theta Nfloors}$	$I_{T\_\theta Nfloors}$	$I_{M2\_\theta Nfloors}$	$I_{M3\_\theta Nfloors}$
Mid section 3 Nfloors						
Terminal section 3 Nfloors						

→ The variable **dif** (540, 6, 7800) is saved in dif_all.mat. The variable contains 12 types of columns and beams, respectively. The units used in this example are N for force and N·m for moment.

Note: The three sections of a structural member considered in the influence coefficients are not necessarily the member's ends and midpoint. They should include the critical sections defined in accordance with specifications in ACI 318, which may occur at other locations.

Dynamic properties	
No. of modes	= Number of vibrational modes to be considered in the analysis

	$\rightarrow$ 3 (modes)
Modal Periods	= Modal periods of the vibrational modes (s) $\rightarrow$ [6.05 5.70 5.00] (sec) Note: Multiple variables must be input using square brackets as shown
Mode shapes	= Load the MATLAB file (.mat) containing the mode shapes The variable is named "**evectors**" and can be saved in a mat file with an arbitrary name. Number of column vectors = number of mode shapes (lowest from the left). Each column has the x-coordinates, y-coordinates, and then θ-coordinates. **evectors** (Fdofs*Nfloors, No of modes) = $$\begin{Bmatrix} \phi_{x1} \\ \phi_{x2} \\ \vdots \\ \phi_{xN} \\ \hdashline \phi_{y1} \\ \phi_{y2} \\ \vdots \\ \phi_{yN} \\ \hdashline \phi_{\theta 1} \\ \phi_{\theta 2} \\ \vdots \\ \phi_{\theta N} \end{Bmatrix}_{\text{1st mode}} \begin{Bmatrix} \phi_{x1} \\ \phi_{x2} \\ \vdots \\ \phi_{xN} \\ \hdashline \phi_{y1} \\ \phi_{y2} \\ \vdots \\ \phi_{yN} \\ \hdashline \phi_{\theta 1} \\ \phi_{\theta 2} \\ \vdots \\ \phi_{\theta N} \end{Bmatrix}_{\text{2nd mode}} \begin{Bmatrix} \phi_{x1} \\ \phi_{x2} \\ \vdots \\ \phi_{xN} \\ \hdashline \phi_{y1} \\ \phi_{y2} \\ \vdots \\ \phi_{yN} \\ \hdashline \phi_{\theta 1} \\ \phi_{\theta 2} \\ \vdots \\ \phi_{\theta N} \end{Bmatrix}_{\text{3rd mode}}$$ $\rightarrow$ evectors (180, 3) is saved in ModeShapes.mat
Modal damping ratio	= Modal damping as a percentage of critical damping (%) $\rightarrow$ [2 2 2] (%)
Mass	= Load the MATLAB file containing the mass and mass moment of inertia associated with each floor The variable is named "**mass**" and can be saved in a mat file with an arbitrary name. The variable mass is a column vector. The first 3 rows are associated with the mass in the x-direction and in the y-direction (M_x and M_y), and mass moment of inertia in the θ-direction (M_θ) of the first floor. The next 3 rows are associated with the second floor and so forth. mass (3×Nfloors, 1) =

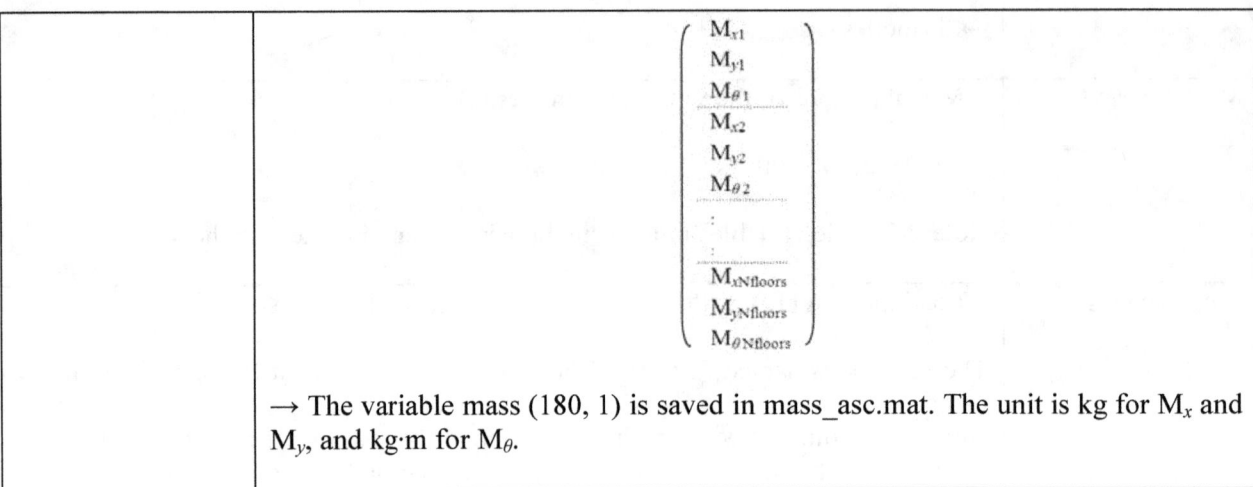

$$\begin{pmatrix} M_{x1} \\ M_{y1} \\ M_{\theta 1} \\ \hline M_{x2} \\ M_{y2} \\ M_{\theta 2} \\ \vdots \\ \vdots \\ \hline M_{xNfloors} \\ M_{yNfloors} \\ M_{\theta Nfloors} \end{pmatrix}$$

$\rightarrow$ The variable mass (180, 1) is saved in mass_asc.mat. The unit is kg for M_x and M_y, and kg·m for M_{θ}.

Load Modeling:

Load Modeling				
Load factors	Dead loads: 1.2	Super-imposed dead loads: 1.2	Live loads: 1	Wind loads: 1
Gravity loads	Specify dead loads on prototype floors [N]: C:\HR_DAD_RC\Input\frames_DeadLoad.mat [Select file...]		Specify super-imposed dead loads on prototype floors [N]: C:\HR_DAD_RC\Input\frames_SDeadLoad.mat [Select file...]	
	Specify live loads on prototype floors [N]: C:\HR_DAD_RC\Input\frames_LiveLoad.mat [Select file...]			
Wind loads	Specify time-history wind loads in a model scale [N, N.m]: C:\HR_DAD_RC\fl_suburban\Fl_xxx [Select file...]			
Wind tunnel test data	Mean wind speed [m/s]: 23.2	Wind directions [deg.]: 0 10 20 30 40 50 60 7	Model scale [/1]: 500	
	Sampling rate [Hz]: 250	No. of sampling points: 7504	Threshold point: 200	
Speed range	Wind speeds for response databases [m/s]: 20 30 40 50 60 70 80			
Lower limit requirement	☑ ASCE 7-based overturning moments [N.m]: C:\HR_DAD_RC\Input\moment_ovtn_ASCE.mat [Select file...]			

Load factors	
Dead loads	= Dead load factor Load factor of dead loads in a load combination of factored loads. → 1.2 (from the load combination 1.2D + 1.0L + 1.6W using ASCE 7-05 Standards)
Super-imposed dead load	= Super-imposed Dead load factor Load factor of super-imposed dead loads in a load combination of factored loads. → 1.2 (from the load combination 1.2D + 1.0L + 1.6W using ASCE 7-05 Standards)
Live loads	= Live load factor Load factor of live loads in a load combination of factored loads. → 1.0 (from the load combination 1.2D + 1.0L + 1.6W using ASCE 7-05 Standards)
Wind loads	= Wind load factor Load factor of wind loads in a load combination of factored loads. → 1.0 (from the load combination 1.2D + 1.0L + 1.6W using ASCE 7-05 Standards) Note: Because HR_DAD accounts for wind effects for specified MRIs appropriate for the limit state of interest, the wind load factor being used is not 1.6 but 1.0.

Gravity loads	
Specify dead loads on prototype floors	= Load the MATLAB file containing the contribution of the dead loads The variable is named "**frames_DL**" and can be saved in a mat file with an arbitrary name. The variable **frames_DL** is a matrix with the first column of member IDs that coincide with the first row of **mem_list**. Each row contains the internal forces due to dead loads occurring in the initial, mid, and terminal sections of the member identified in the first column of the row. All 6 internal forces and moments (P, V2, V3, T, M2, and M3) are assigned from columns 2 to 7 for the initial section, from columns 8 to 13 for the mid section, and from columns 14 to 19 for the terminal section. **frames_DL** (No. of members (=Nm), 19) =

$$
\begin{array}{cccc}
\begin{array}{c}\text{Member} \\ \text{ID}\end{array} & \begin{array}{c}\text{Initial} \\ \text{section}\end{array} & \begin{array}{c}\text{Mid} \\ \text{section}\end{array} & \begin{array}{c}\text{Terminal} \\ \text{section}\end{array}
\end{array}
$$

$$
\left[
\begin{array}{c|cccccc|c|c}
1 & P_{D1} & V2_{D1} & V3_{D1} & T_{D1} & M2_{D1} & M3_{D1} & & \\
2 & P_{D2} & V2_{D2} & V3_{D2} & T_{D2} & M2_{D2} & M3_{D2} & & \\
\vdots & \vdots & \vdots & \vdots & \vdots & \vdots & \vdots & & \\
Nm & P_{DNm} & V2_{DNm} & V3_{DNm} & T_{DNm} & M2_{DNm} & M3_{DNm} & &
\end{array}
\right]
$$

	→ The variable **frames_DL** (7800, 19) is saved in frames_DeadLoad.mat. The units are N for forces and N·m for moments. Note: The forces and moments of **frames_DL** are due to unfactored dead loads. They can be calculated using a finite element analysis software.
Specify super-imposed dead loads on prototype floors	= Load the MATLAB file containing the contribution of the super-imposed dead loads The variable is named "**frames_SDL**" and can be saved in a mat file with an arbitrary name. The format of **frames_SDL** is consistent with that of **frames_DL**. **frames_SDL** (No. of members (=Nm), 19) =

$$
\begin{array}{cccc}
\begin{array}{c}\text{Member} \\ \text{ID}\end{array} & \begin{array}{c}\text{Initial} \\ \text{section}\end{array} & \begin{array}{c}\text{Mid} \\ \text{section}\end{array} & \begin{array}{c}\text{Terminal} \\ \text{section}\end{array}
\end{array}
$$

$$
\left[
\begin{array}{c|cccccc|c|c}
1 & P_{S1} & V2_{S1} & V3_{S1} & T_{S1} & M2_{S1} & M3_{S1} & & \\
2 & P_{S2} & V2_{S2} & V3_{S2} & T_{S2} & M2_{S2} & M3_{S2} & & \\
\vdots & \vdots & \vdots & \vdots & \vdots & \vdots & \vdots & & \\
Nm & P_{SNm} & V2_{SNm} & V3_{SNm} & T_{SNm} & M2_{SNm} & M3_{SNm} & &
\end{array}
\right]
$$

	→ The variable **frames_SDL** (7800, 19) is saved in frames_SDeadLoad.mat. The units are N for forces and N·m for moments. Note: The forces and moments of **frames_DL** are due to unfactored super-imposed dead loads. They can be calculated using a finite element analysis software.
Specify live loads	= Load the MATLAB file containing the contribution of the live loads

on prototype floors	The variable is named "**frames_LL**" and can be saved in a mat file with an arbitrary name. The format of **frames_SDL** is consistent with that of **frames_DL**.

frames_SDL (No. of members (=Nm), 19) =

Member ID	Initial section						Mid section	Terminal section
1	P_{L1}	$V2_{L1}$	$V3_{L1}$	T_{L1}	$M2_{L1}$	$M3_{L1}$		
2	P_{L2}	$V2_{L2}$	$V3_{L2}$	T_{L2}	$M2_{L2}$	$M3_{L2}$		
⋮	⋮	⋮	⋮	⋮	⋮	⋮		
Nm	P_{LNm}	$V2_{LNm}$	$V3_{LNm}$	T_{LNm}	$M2_{LNm}$	$M3_{LNm}$		

→ The variable **frames_LL** (7800, 19) is saved in frames_LiveLoad.mat. The units are N for forces and N·m for moments.

Note: The forces and moments of **frames_LL** are due to unfactored live loads. They can be calculated using a finite element analysis software.

Wind loads	
Specify time-history wind loads in a model scale	= Load the MATLAB file containing the time histories of the floor loads of a model in wind tunnel tests

The variable is named "F" and can be saved under any name but must end with "_XXX". The subfix XXX gives the direction of wind θ_w in degrees in wind tunnel. An example is shown in Fig. 2 for wind directions XXX = 000 (0°) and XXX = 045 (45°).

The variable contains time-history of a floor load in one of x, y, θ directions. The first Nfloors (i.e., the number of floors) rows correspond to the floor loads acting in the x-direction starting from the first floor. The next Nfloors and the last Nfloors rows correspond to ones in the y-direction and in the θ-direction, respectively.

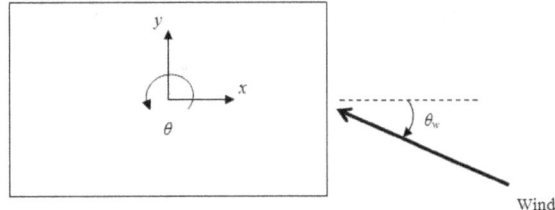

F(Fdof*Nfloors, Npoints) =

$$\begin{pmatrix} Fx_{1\_t=1} & Fx_{1\_t=2} & \cdots & Fx_{1\_t=Npoints} \\ Fx_{2\_t=1} & Fx_{2\_t=2} & \cdots & Fx_{2\_t=Npoints} \\ \vdots & \vdots & & \vdots \\ Fx_{Nfloors\_t=1} & Fx_{Nfloors\_t=2} & \cdots & Fx_{Nfloors\_t=Npoints} \\ Fy_{1\_t=1} & Fy_{1\_t=2} & \cdots & Fy_{1\_t=Npoints} \\ Fy_{2\_t=1} & Fy_{2\_t=2} & \cdots & Fy_{2\_t=Npoints} \\ \vdots & \vdots & & \vdots \\ Fy_{Nfloors\_t=1} & Fy_{Nfloors\_t=2} & \cdots & Fy_{Nfloors\_t=Npoints} \\ F\theta_{1\_t=1} & F\theta_{1\_t=2} & \cdots & F\theta_{1\_t=Npoints} \\ F\theta_{2\_t=1} & F\theta_{2\_t=2} & \cdots & F\theta_{2\_t=Npoints} \\ \vdots & \vdots & & \vdots \\ F\theta_{Nfloors\_t=1} & F\theta_{Nfloors\_t=2} & \cdots & F\theta_{Nfloors\_t=Npoints} \end{pmatrix}$$

$\rightarrow$ F(180, 7504) is saved in Fl_XXX files where XXX is from 000 to 360 by 10 increment. The unit is N for force and N.m for moment.

Note: Terrain conditions for the wind tunnel model must simulate prototype conditions.

Wind tunnel test data	
Mean wind speed	= Hourly mean wind speed at roof of the model during wind tunnel tests (m/s) $\rightarrow$ 23.2 m/s for suburban terrain Note: This speed is an hourly mean speed.
Wind directions	= Wind directions used in the wind tunnel test The variable **WD** can be a vector containing clockwise directions of wind as shown in Fig. 4. For example, $\theta_w = 0°$ to $360°$ with $10°$ clockwise increments from an arbitrary direction: [0 10 20 … 350 360] $\rightarrow$ [0:10:360] Note 1: The wind directions in **WD** are identical to the wind directions in the wind tunnel tests or CFD simulations. Note 2: The variable can be input using square brackets as shown.

Model scale	= Scale of the model used in the wind tunnel tests If the scale of the model to the prototype structure is 1/500, the value is 500. → 500
Sampling rate	= Sampling frequency used in the wind tunnel tests (Hz) → 250 Hz
No. of sampling points	= Total number of points that make up the time histories of the floor loads → 7504 (from data of 30 s with a sampling frequency of 250 Hz in the wind tunnel)
Threshold point	= Number of points to be cut from beginning of the time histories during the analysis → 200 Note: Numerical integration needs a certain number of points before it stabilizes. Therefore a certain number of initial points must be cut from the solution of the equations of motion before estimating the response parameters.
Speed range	
Wind speeds for response databases	= One-hour mean wind speeds at roof height of the full scale building for which the response in desired The variable **WS** is a vector containing wind speeds directed toward the structure. For example, 20 m/s to 80 m/s (one-hour mean wind speeds) in 10 m/s increments: [20 30 40 50 60 70 80] → [20:10:80] Note: The variable can be input using square brackets as shown. The wind speed range must cover wind speeds at the roof height based on wind climatological data (e.g., hurricane data) at the site and must account for the building orientation.
Lower limit requirement	
ASCE 7-based overturning moments	= ASCE 7-based overturning moments in the principal axes corresponding to Risk Category of building According to ASCE 7-05, Section C6.6, forces and pressures estimated through wind tunnel testing are to be limited to not less than 80 % of their ASCE 7-based analytical method counterpart. HR_DAD_RC employs overturning moments in the directions of the principal axes as measures of this limitation. Overturning moments based on the ASCE 7 analytical method are based on the importance

factors 0.87 for Risk Category I buildings, 1.0 for Risk Category II buildings, and 1.15 for Risk Category III or IV buildings. These overturning moments are compared to their HR_DAD_RC counterparts.

The variable is named "**Movtn_asce**" and can be saved in a mat file with an arbitrary name. The variable **Movtn_asce** is a matrix in which the first column contains overturning moments along the first principal axis (i.e., x axis) and the second column contains moments along the second principal axis (i.e., y axis). The first row corresponds to ASCE 7-based overturning moments for the first Risk Category being considered for the building. If two or three Risk Categories are considered, they are reflected in the second, or second and third row.

Movtn_asce (No. of MRIs in B_{ij}, No. of principal axes) =

$$
\begin{array}{c}
\text{1}^{\text{st}}\text{ Risk Category} \\
\text{2}^{\text{nd}}\text{ Risk Category} \\
\vdots \\
\text{n}^{\text{th}}\text{ Risk Category}
\end{array}
\begin{array}{cc}
x\text{-axis} & y\text{-axis} \\
\left[\begin{array}{cc}
M_{x1} & M_{y1} \\
M_{x2} & M_{y2} \\
\vdots & \vdots \\
M_{xn} & M_{yn}
\end{array}\right]
\end{array}
$$

$\rightarrow$ **Movtn_asce** (3, 2) is saved in moment_ovtn_ASCE.mat where ASCE 7-based overturning moments in the x and y axes based on three Risk Categories of ASCE 7 correspond to counterpart calculated by HR_DAD_RC with three MRIs specified for demand-to-capacity indexes. The unit is N·m.

Note: Structural designs of buildings with Risk Category I, II, and III or IV in ASCE 7 are assumed to correspond to counterparts from HR_DAD_RC with MRIs of 300 years, 700 years, and 1700 years, respectively (ASCE 2010).

Page 'Wind Effects'

Response Databases:

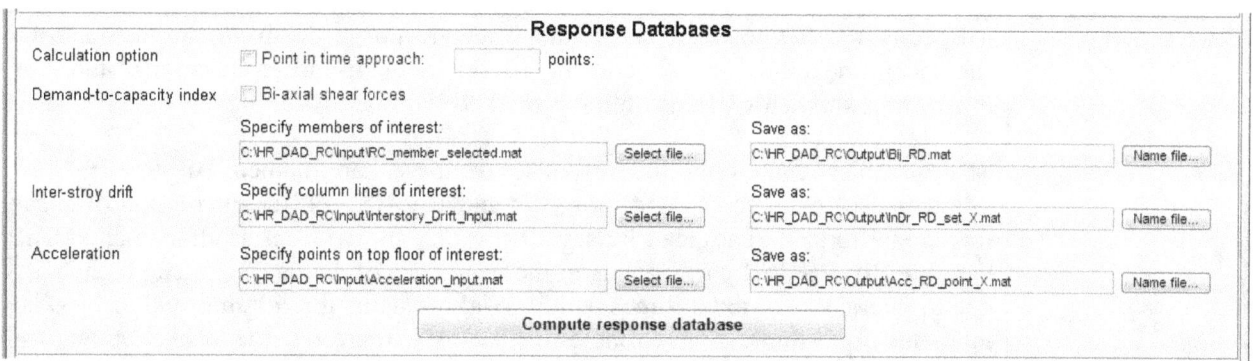

Calculation Option	
Point in time approach	= Check the box if the point in time approach is selected for the calculation of the response databases for the demand-to-capacity indexes. A full time approach for the indexes will be employed unless this option is checked. Note: The point in time approach considerably reduces simulation time.
Points	= Specify the number of points used in point in time approach. Note: If N points are specified, only $7N$ points within the time histories of the internal forces are checked, rather than all the points in those time histories.
Demand-to-capacity index	
Bi-axial shear forces	= Check the box if the demand-to-capacity index for bi-axial shear forces is desired. Note: This option is for columns subjected to bi-axial shear forces.
Specify members of interest	= Load the MATLAB file containing list of members to calculate their demand-to-capacity indexes. The variable is named "**member_selected**" and can be saved in a mat file with an arbitrary name. The variable **member_selected** is a one-row vector that contains a list of the members of interest for which response database for the demand-to-capacity index are calculated. **member_selected** (1, No. of members of interest) = $$\begin{matrix} 1 & 2 & \dots & n & \dots & \text{No. of members of interest} \\ \begin{bmatrix} 1 & 201 & \dots & 301 & \dots & 2000 \end{bmatrix} \end{matrix}$$

	→ **member_selected** (1, 96) is saved in RC_member_selected.mat. The 20 members consisting of the columns and beams of interest were selected for the calculation of the response database for the demand-to-capacity index and peak responses with specified MRIs.
Save as	= Specify the file location and name where response databases for demand-to-capacity indexes of the members (defined in **member_selected**) and for overturning moments (if this option is clicked) will be saved. The variables containing the response databases are named **Bij_RS_PM** for interaction of axial forces and moments and **Bij_RS_VT** for interaction of shear forces and torsion. They can be saved in a mat file with an arbitrary name. The variable **Bij_RS_PM** corresponds to the peak ratio of demand to strength capacity due to interaction of axial forces and bi-axial moments for columns and of bi-axial moments for beams. The variable **Bij_RS_VT** represent the peak demand-to-capacity indexes of all members due to due to interaction of shear forces and torsion. The variables are 3D arrays with the same format, each face of which represents the response databases for demand-to-capacity indexes of a specific member. The index of each face corresponds to the index of selected members defined in "**member_selected**". An element (i, j) of the pth face corresponds to the peak demand-to-capacity index of the member p for the wind with the ith wind direction of **WD** and the jth wind speed of **WS**. For example, a value at element (2, 5, 38) of **Bij_RS_PM** represents the peak demand-to-capacity index for axial forces and moments for the 38th member in the list of **member_selected** in a given wind of the 2nd wind direction of 10° in **WD** and the 5th wind speed of 60 m/s in **WS**. **Bij_RS_PM** (No. of WD, No. of WS, No. of selected members) = 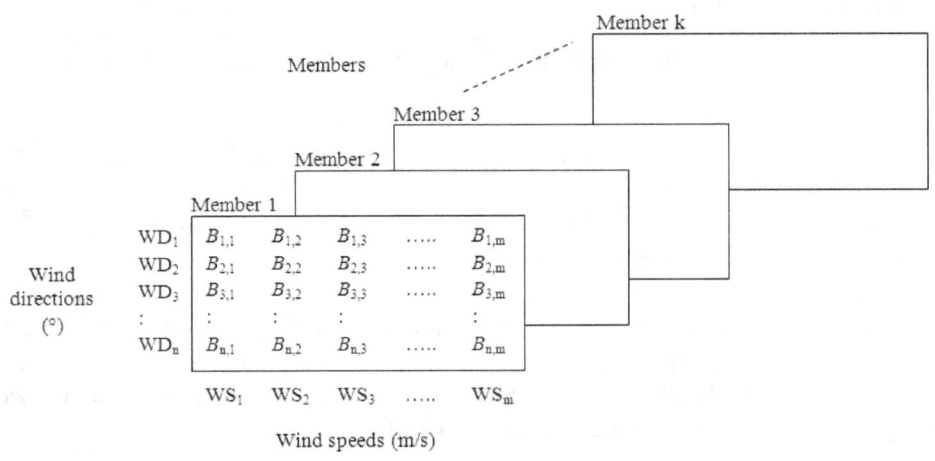 If the option of the lower limit requirement for the overturning moments is selected in the section 'Load Modeling', page 'Modeling', the additional variables containing the response databases for the overturning moments are named **Mx_ovtn** and **My_ovtn** for overturning moments in the x and y principal axes,

A27

respectively. The variables have the same format as the format for **Bij_RS_PM**, except that for the overturning moment there is only one face. An element (i, j) corresponds to the overturning moment for the wind with i^{th} direction **WD** and j^{th} speed **WS**.

→ **Bij_RS_PM** (37, 7, 96), **Bij_RS_VT** (37, 7, 96), **Mx_ovtn** (37, 7), and **My_ovtn** (37, 7) are saved in Bij_RD.mat.

Note: If the variable is higher than unity the strength capacity of the member is not adequate for axial forces and moments (**Bij_RS_PM**) or for shear forces and torsional moment (**Bij_RS_VT**). The corresponding reinforcement and/or dimension of the member must be modified in order to make the variables less than or equal to unity.

Inter-story drift	
Specify column lines of interest	= Load the MATLAB file containing the position of the column lines where the response database of inter-story drift is desired. The variable is named "**interstory_location**" and can be saved in a mat file with an arbitrary name. The variable is a matrix where the first Nfloors rows contain the x and y coordinates of the column lines with respect to the mass center of each floor in the first and the second column, and the height of the story is assigned in the third column. The first row contains information relating to the first floor, the second row stores that of the second floor, and so forth. Successive column lines are appended as an extra Nfloors rows (where N is the number of floors). **interstory_location** (Nc×Nfloors, 3) for Nc column lines = $$\begin{matrix} & & x & y & \text{Story} \\ & & \text{coord.} & \text{coord.} & \text{height} \end{matrix}$$ $$\begin{matrix}\text{Column}\\\text{line 1}\end{matrix}\left\{\begin{matrix} x_{1\_L1} & y_{1\_L1} & h_{1\_L1} \\ x_{2\_L1} & y_{2\_L1} & h_{2\_L1} \\ \vdots & \vdots & \vdots \\ x_{N\_L1} & y_{N\_L1} & h_{N\_L1} \end{matrix}\right\}$$ $$\begin{matrix}\text{Column}\\\text{line Nc}\end{matrix}\left\{\begin{matrix} x_{1\_LNc} & y_{1\_LNc} & h_{1\_LNc} \\ x_{2\_LNc} & y_{2\_LNc} & h_{2\_LNc} \\ \vdots & \vdots & \vdots \\ x_{N\_LNc} & y_{N\_LNc} & h_{N\_LNc} \end{matrix}\right\}$$ → **interstory_location** (120,3) for 2 column lines is saved in Interstory_Drift_input.mat.
Save as	= Specify the file location and name where the response databases for inter-story drift will be saved. The variables containing the response databases are named **InDr_RS_set_X**, where X depends on the column line defined in **interstory_location**, and can be saved in a mat file with an arbitrary name.

The variable **InDr_RS_set_X** is a 3D array, each face of which represents the response surface in a specific direction (x or y). The first Nfloor faces (Nfloor is the number of floors of the structure) of the array are associated with the response in the x direction while the next Nfloor faces are associated with the y direction.

An element (i, j) of a specific face in a given set X corresponds to the peak inter-story drift of the column lines considered in set X for the wind with i^{th} direction **WD** and j^{th} speed **WS**. The face represents the floor to be considered inter-story drift in x or y direction. The details are provided in the following schematic.

InDr_RS_set_X (No. of WD, No. of WS, 2×Nfloors) =

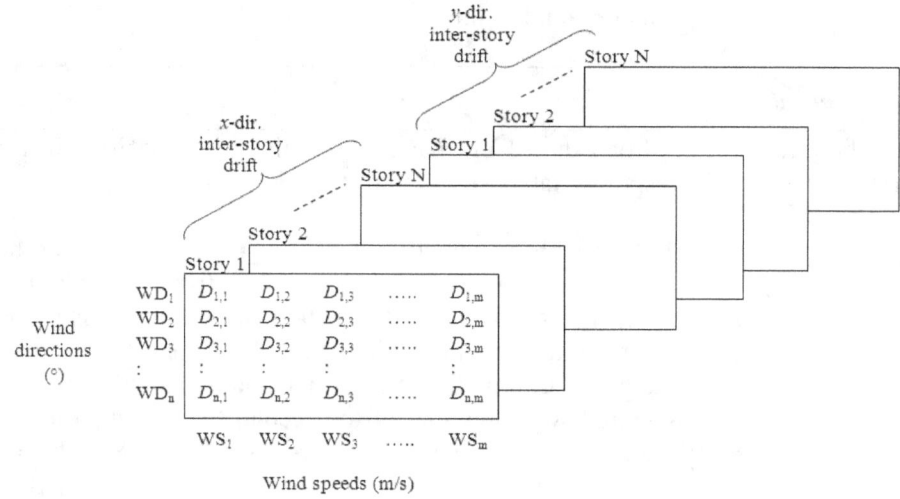

→ **InDr_RD_set_1** (5, 37, 120) and **InDr_RD_set_2** (5, 37, 120) for two column lines are saved in InDr_RS_set_1.mat and InDr_set_2.mat, respectively.

Acceleration

| Specify points on top floor of interest | = Load the MATLAB file containing the positions of the points belonging to the top floor where the response database of peak acceleration is desired.

The variable is named "**acceleration_location**" and can be saved in a mat file with an arbitrary name. The variable is a matrix where each row contains the x and y coordinates of a point with respect to the mass center of the top floor. The first row contains information relating to the first point, the second row stores information for the second point, and so forth. A successive point is appended as an extra row.

acceleration_location (Np, 2) for N points = |

$$
\begin{array}{c}
 \\
\text{Point 1} \\
\text{Point 2} \\
 \\
\text{Point N}
\end{array}
\begin{array}{cc}
x & y \\
\text{coord.} & \text{coord.} \\
\left(\begin{array}{cc} x_{P1} & y_{P1} \\ x_{P2} & y_{P2} \\ \vdots & \vdots \\ x_{PN} & y_{PN} \end{array} \right)
\end{array}
$$

	→ The variable **acceleration_position** (4, 2) for 4 points is saved in Acceleration_input.mat.
Save as	= Specify the file location and name where the top floor acceleration response databases will be saved. The variable containing the response databases is named **Acc_RS_point_X** where X depends on the point belonging to the top floor defined in **acceleration_location**, and can be saved in a mat file with an arbitrary name. The variable **Acc_RS_point_X** is a 3D array, each face of which represents the response surface in a specific direction (x or y). The first face of the array is associated with the response in the x direction, and the next face is associated with the y direction. An element (i, j) of a specific face in a given point X corresponds to the peak acceleration of the top floor considered at the point X for the wind with i^{th} direction **WD** and j^{th} speed **WS**. **Acc_RS_point_X** (No. of WD, No. of WS, 2) =

$$
\begin{array}{c}
\text{y-dir} \\
\text{x-dir.} \\
\begin{array}{c}
\text{Wind} \\
\text{directions} \\
(^\circ)
\end{array}
\begin{array}{c}
WD_1 \\ WD_2 \\ WD_3 \\ \vdots \\ WD_n
\end{array}
\left|
\begin{array}{ccccc}
A_{1,1} & A_{1,2} & A_{1,3} & \cdots & A_{1,m} \\
A_{2,1} & A_{2,2} & A_{2,3} & \cdots & A_{2,m} \\
A_{3,1} & A_{3,2} & A_{3,3} & \cdots & A_{3,m} \\
\vdots & \vdots & \vdots & & \vdots \\
A_{n,1} & A_{n,2} & A_{n,3} & \cdots & A_{n,m}
\end{array}
\right| \\
\begin{array}{ccccc}
WS_1 & WS_2 & WS_3 & \cdots & WS_m
\end{array} \\
\text{Wind speeds (m/s)}
\end{array}
$$

	→ **Acc_RD_point_1** (5, 37, 120) to **Acc_RD_point_4** (5, 37, 120) for four column lines are saved in Acc_RS_point_1.mat to Acc_RS_point_4.mat.
"Compute response database"	
	Click the button, and the response databases defined above are calculated. Note: The calculation can be performed when all variables are provided correctly in the sections 'Building Modeling', 'Load Modeling', and 'Response Databases'.

A30

Peak Responses for specified MRIs:

Wind climatological data	**Directory containing simulated hurricane wind speed files:**	**Select the milepost corresponding to the building location:**
	C:\HR_DAD_RC\hurricane_dataset [Select dir...]	Milepost 1400 / Milepost 1450 / Milepost 1500 [Check]
	Terrain exposure at the weather **Threshold wind speed [m/s]:**	
	C 20	
Micro-meteorological data	1) ASCE 7 ⦿ or 2) User-defined	[Select file...]
Terrain data	**Terrain exposure surrounding the building:** C:\HR_DAD_RC\Input\terrain_exposure.mat [Select file...]	**Building orientation angle [deg.]:** 90
Calculation option	☐ Veering angle effects:	
Peak responses for MRIs	**MRIs for demand-to capacity index [years]:** 300 700 1700	**Save as:** C:\HR_DAD_RC\Output\Peak_Bij.mat [Name file...]
	MRIs for inter-story drift [years]: 20	**Save as:** C:\HR_DAD_RC\Output\Peak_InDr_set_X.mat [Name file...]
	MRIs for acceleration [years]: 10	**Save as:** C:\HR_DAD_RC\Output\Peak_Acc_point_X.mat [Name file...]

[Compute peak response with specified MRIs]

Wind climatological data

Note: This module needs to be modified depending upon the wind climatological database being used.

The following pertains to the simulated 999 extreme wind events provided for a large number of locations (mileposts) along the Gulf of Mexico and North Atlantic coast. This simulated data is publicly available at www.nist.gov/wind by following the links for extreme wind data sets.

The database sets are one-minute mean hurricane wind speeds in knots at 10m above the ground in open terrain near the coast line. The wind directions are from 22.5° to 360° with 22.5° clockwise increments from the North.

Directory containing simulated hurricane wind speed files	= Specify the folder location of the database of simulated hurricanes → c:\HR_DAD_RC\hurricanes_dataset Note: The hurricanes database can be downloaded at www.nist.gov/wind.
Select the milepost corresponding to the building location	= Hurricane milepost that can represent the building location → "Milepost 1450" for a building located in Miami, FL. Note: When the milepost is selected for a specific building location in the listbox, the "Check" button must be pushed for confirmation.
Terrain exp. at the weather station	= Terrain exposure at weather station where hurricane winds are measured. The terrain exposure is categorized as "B", "C", and "D" for suburban, open, and unobstructed terrains, respectively, as defined in ASCE 7-05.

A31

	→ "C" for open terrain in all directions Note: Terrain exposure of weather stations is assumed to be identical in all directions.
Min. wind speeds for calculation	= Minimum one-hour wind speed under which the response is no longer of interest → 15 m/s Note: This speed is higher than or equal to the minimum wind speed of WS defined in Page Three. A low minimum wind speed will lengthen the calculation of the responses with a specified MRI. However, the higher the minimum wind speed, the greater is the possibility of not accounting for critical wind effects, especially for serviceability requirements.

Micro-meteorological data

ASCE 7	= Check the box if the ASCE 7-based calculation is used in predicting the micro-meteorological relationship of wind speeds at the weather station, e.g., at 10 m above the ground in open terrain exposure, to speeds at the rooftop of the building for the appropriate terrain exposures.
User-defined	= Specify a filename containing the micro-meteorological relationship defined by a user. The variable must be named "**Ratio_Vs**" and can be saved in a mat file with an arbitrary name. The variable **Ratio_Vs** can be a vector containing the ratios between wind speeds at the weather station to the mean hourly wind speeds at the building height for the requisite terrain exposures. The current version of HR_DAD_RC uses wind speeds of simulated 999 extreme wind events for a large number of locations (mileposts) along the Gulf of Mexico and North Atlantic coast (www.nist.gov/wind). **Ratio_Vs** (1, 4 (= total number of terrain exposure conditions)) = $$\text{Terrain exposure:} \quad A \qquad B \qquad C \qquad D$$ $$\left[\frac{V_A^H}{V_C^{10m}}, \quad \frac{V_B^H}{V_C^{10m}}, \quad \frac{V_C^H}{V_C^{10m}}, \quad \frac{V_D^H}{V_C^{10m}} \right]$$ where V in the numerator is a wind speed at building height (H) corresponding to each terrain exposure condition from A to D, and V in the denominator is a wind speed at a weather station at 10 m above ground in open (C) terrain. → **Ratio_Vs** (1, 4) is saved in micro-meteor.mat; this example uses ASCE 7-based micro-meteorological data by checking the ASCE 7 box above Note: The micro-meteorological data must be provided by an expert in wind engineering.

Terrain data	
Terrain exposure surrounding the building	= Terrain exposure surrounding the building in each direction. The variable is named "**terrain**" and can be saved in a mat file with an arbitrary name. The variable terrain can be a vector containing terrain roughness in 16 directions clockwise from 22.5° to 360° with 22.5° increments from the North. The directions of terrain exposure are identical to those of hurricane winds described in the "Wind climatological data" section. The terrain exposure is categorized as "B", "C", and "D" for suburban, open, and unobstructed terrains, respectively, as defined in ASCE7-05. The terrains exposure can be different according to directions, which enables HR_DAD to account for directionality effects of the terrain exposure. **terrain** (1, 16) = $$\begin{matrix} 1 & 2 & .... & 15 & 16 \\ ["C" & "C" & .... & "B" & "B"] \end{matrix}$$ → ["B" "B" .... "B"] for suburban terrain in all directions and is saved in terrain_exposure.mat Note: The terrain exposure surrounding the building must be identical to that used in wind tunnel tests for the aerodynamic database described in the "Wind loads" section.
Building orientation	= Orientation angle (α_0) of a building in the clockwise direction from the North to the x axis of the building ($0° \leq \alpha_0 < 360°$). 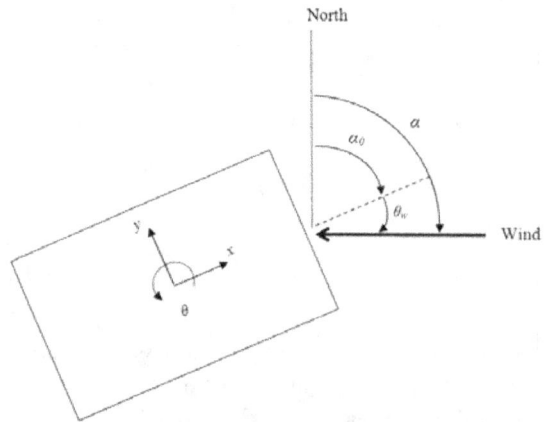 → 90 [degrees] Note: If the terrain exposure does not change around a building, the orientation angle of the building can be changed. However, for direction-dependent terrain conditions, various orientation angles can be taken into account only when aerodynamic data from wind tunnel tests account for terrain exposure

	corresponding to the various building orientations of interest.
Calculation option	
Veering angle effects	= Check the box if veering angle of wind is taken into account. Note: Wind directions at the top of a high-rise building are different from those at 10 m above ground due to veering (change of wind direction with elevation). Clicking the button activates the HR_DAD_RC veering angle option.
Peak responses with MRIs	
MRIs for demand-to-capacity index	= Mean Recurrence Intervals for demand-to-capacity index of selected members → [300 700 1700] Note: The unit is year.
Save as	= Specify the file location and name where the peak response of demand-to-capacity indexes for selected members with specified MRIs will be saved. The variables containing the selected members' demand-to-capacity indexes for axial force and bending moments and for shear forces and torsional moment with specified MRIs are named **Bij_PM_MRIs** and **Bij_VT_MRIs**, respectively, and can be saved in a mat file with an arbitrary name. The variables **Bij_PM_MRIs** and **Bij_VT_MRIs** are matrices the rows of which contain the selected members' peak demand-to-capacity indexes (B_{ij}) for the MRI specified above. The members of interest are defined in "**member_selected**" in the 'Response Databases' section. The first row will contain the indexes corresponding to the first MRIs for demand-to-capacity index defined above, the second row will have those corresponding to the second MRI, and so forth. The index of each column is the index of the members defined in **member_selected**. For example, an element (3, 300) of **Bij_PM_MRIs** represents the peak demand-to-capacity index for axial force and bending moments of the 300th member in the list **member_selected** with the 3rd of the MRIs specified for the demand-to-capacity index. **Bij_PM_MRIs** (No. of MRIs, No. of selected members (=Nm)) = → **Bij_PM_MRIs** (3, 96) and **Bij_VT_MRIs** (3, 96) of 96 selected members for three MRIs are saved in Peak_Bij.mat. Note 1: If the option of the lower limit requirement for the overturning moment is selected in the section 'Load Modeling' of the page 'Modeling', the variables **Scaled_Bij_PM_MRIs** and **Scaled_Bij_VT_MRIs** containing the adjusted peak

A34

	demand-to-capacity indexes for the selected members are also saved.
	Note 2: Additional variables saved in the file can be used for post analysis:
	- **Mx_ovtn_MRIs** (No. of MRIs, 1): Overturning moments along the x axis corresponding to the specified MRIs.
	- **My_ovtn_MRIs** (No. of MRIs, 1): Overturning moments along the y axis corresponding to the specified MRIs.
	- **Rt_Mx_ovtn** (No. of MRIs, 1): Ratio of x-axis overturning moments calculated by DAD to x-axis overturning moments by ASCE 7 corresponding to the specified MRIs.
	- **Rt_My_ovtn** (No. of MRIs, 1): Ratio of y-axis overturning moments calculated by DAD to y-axis overturning moments by ASCE 7 corresponding to the specified MRIs.
	- **Scale_Mx_ovtn** (No. of MRIs, 1): Index adjustment factor for the x axis corresponding to the specified MRIs.
	- **Scale_My_ovtn** (No. of MRIs, 1): Index adjustment factor for the y axis corresponding to the specified MRIs.
	- **Scale_M_ovtn** (No. of MRIs, 1): Index adjustment factor for the x- or y-axis, whichever is larger.
	- **MRI_sorted_Bij** (No. of hurricane events, 1) MRI corresponding to the rank order of the hurricane events (in descending ranking order).
	- **sorted_Bij_PM** (No. of hurricane events, No. of selected members) Selected members' peak demand-to-capacity index for axial force and bending moments corresponding to all hurricane events considered in descending ranking order.
	- **sorted_Bij_VT** (No. of hurricane events, No. of selected members) Selected members' peak demand-to-capacity index for shear forces and torsional moment corresponding to all hurricane events considered in descending ranking order.
	Note 3: If the veering effect is considered, variables described above with the suffix "_vr" are added in the resulting file, e.g., **Bij_PM_MRIs_vr** and **Bij_VT_MRIs_vr**.
MRIs for inter-story drifts	= Mean Recurrence Intervals for inter-story drift of selected column lines → [20] Note: The unit is year.

Save as	= Specify the file location and name where the peak responses of inter-story drift with specified Mean Recurrence Intervals (MRIs) are saved. The variables containing the inter-story responses with specified MRIs are named **InDr_MRIs_set_X** where X depends on the column line defined in **interstory_location** (in the 'Response Databases' section), and can be saved in a mat file with an arbitrary name. The variable **InDr_MRIs_set_X** is a 3D array, each face of which represents the inter-story drift of a given column line in x and y directions and the resultant drift with specified MRIs. The first face of the array is associated with the response of column line X corresponding to the first MRI, the second face corresponds to the second MRI, and so forth. Each face has two columns corresponding to the x and y directions, respectively, and N rows that correspond to the N floors. **InDr_MRI_set_X** (N, 3, No. of MRIs) for a given set X = 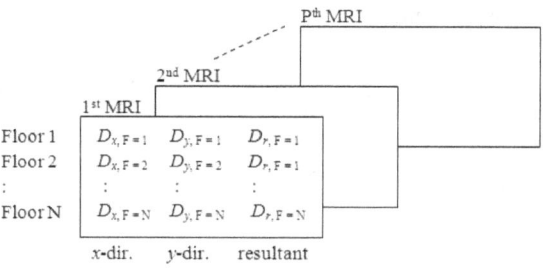 → The variables **InDr_MRI_set_1** (60, 2, 1) to **InDr_MRI_set_4** (60, 2, 1) for 4 sets with 1 specified MRI are saved in Peak_InDr_set_1.mat to Peak_InDr_set_4, respectively. Note 1: Additional variables saved in the file can be used for post analysis: - **sorted_peak_drift_x_set_X** (No. of hurricane events, Number of stories) Peak x-axis inter-story drift ratio for all stories in the set X corresponding to all hurricane events considered in descending ranking order. - **sorted_peak_drift_y_set_X** (No. of hurricane events, Number of stories) Peak y-axis inter-story drift ratio for all stories in the set X corresponding to all hurricane events considered in descending ranking order. - **sorted_peak_drift_t_set_X** (No. of hurricane events, Number of stories) Resultant peak inter-story drift ratio for all stories in the set X corresponding to all hurricane events considered in descending ranking order. Note 2: In the case considering the veering effect, variables described above with suffix of "**_vr**" are added before "**_set_X**", e.g., **InDr_MRI_vr_set_X**.
MRIs for	= Mean Recurrence Intervals for accelerations of selected points on top floor

| top-floor accelerations | → [10]

Note: The unit is year. |
|---|---|
| Save as | = Specify the file location and name where the accelerations of the top floor with specified Mean Recurrence Intervals (MRIs) are saved.

The variables containing the top floor acceleration responses with specified MRIs are named **Acc_MRIs_point_X** where X depends on the point belonging to the top floor defined in **acceleration_location** (in the 'Response Databases' section), and can be saved in a mat file with an arbitrary name.

The variable **Acc_MRIs_point_X** is a 3D array, each face of which represents the accelerations of the top floor at the point X in the x and y directions and the resultant acceleration with specified MRIs. The first face of the array is associated with the response of the point X corresponding to the first MRI, the second face corresponds to the second MRI, and so forth. Each face has three columns corresponding to the x and y directions and the resultant.

Acc_MRIs_point_X (1, 3, No. of MRIs) for a given point X =

→ The variables **Acc_MRIs_point_1** (1, 3, 1) to **Acc_MRIs_point_4** (1, 3, 1) for 4 points with one specified MRI of 10 years are saved in from Peak_Acc_point_1.mat to Peak_Acc_point_4.

Note 1: Additional variables saved in the file can be obtained for post analysis:

- **sorted_peak_Acc_x_point_X** (No. of hurricane events, 1)
Peak x-axis acceleration for point X corresponding to all hurricane events considered in descending ranking order.

- **sorted_peak_Acc_y_point_X** (No. of hurricane events, 1)
Peak y-axis acceleration for point X corresponding to all hurricane events considered in descending ranking order.

- **sorted_peak_Acc_t_point_X** (No. of hurricane events, 1)
Resultant peak acceleration for point X corresponding to all hurricane events considered in descending ranking order.

Note 2: In the case considering the veering effect, variables described above with suffix of "**_vr**" are added before "**_point_X**", e.g., **Acc_MRI_vr_point_X_vr**. |

"Computed peak responses with MRIs"	
	Click the button, and the peak responses defined above are calculated. Note: The calculation of peak responses must be executed after the response databases are obtained. If the response databases are not available, calculate the response databases first by executing "Compute response database".

Page 'Results & Plot'

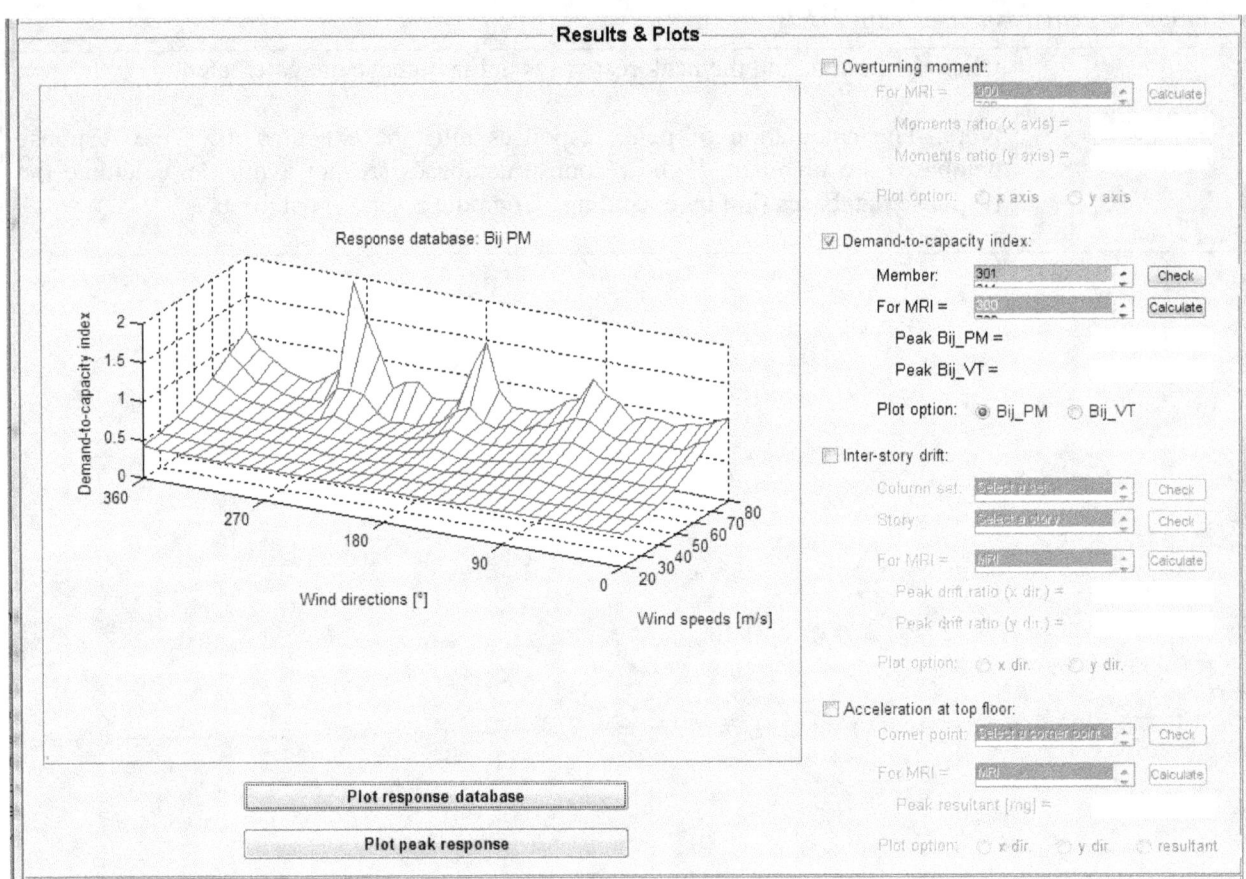

Overturning moments	
Once the button is clicked, the overturning moments module is activated. Note: If the option of lower limit requirement for overturning moments is not clicked in the section 'Load Modeling' of the page 'Modeling', the overturning moments module is not activated.	
For MRI =	= Select a MRI of interest from the list of MRIs, and click the button of "Calculate" → Overturning moments in the x and y axes for the selected MRI will be shown in the blanks. Note: the list of MRIs for overturning moments is obtained from "MRIs for demand-to-capacity index" in the page 'Wind Effects'.
Plot response database	= Place the button of "Plot response database" in the bottom left corner of the page after a plot option (x axis or y axis) is chosen → The response database for the overturning moment in the selected axis will be shown in the left side. The overturning moment is plotted as a function of wind

	speed and direction.
Plot peak response	= Place the button of "Plot response database" in the bottom left corner of the page

→ The peak response database of overturning moments in both x and y axes will be shown along MRIs.

Note: overturning moments for any MRI can be obtained from this plot. |

Demand-to-capacity index

	Once the button is clicked, the demand-to-capacity index module is activated.

| Member | = Select a member of interest from the list of members and click the button of "Check"

Note: the list of members is obtained from "members of interest" in the page 'Wind Effects'. |
|---|---|
| For MRI = | = Select a MRI of interest from the list of MRIs, and click the button of "Calculate"

→ Demand-to-capacity indexes of the selected member for the selected MRI will be shown in the blanks.

Note 1: The list of MRIs for the demand-to-capacity index is obtained from "MRIs for demand-to-capacity index" in the page 'Wind Effects'.

Note 2: Bij_PM denotes the demand-to-capacity index for axial force and moments $\left(B_{ij}^{PM} \right)$. Bij_VT is the demand-to-capacity index for shear forces and torsional moment $\left(B_{ij}^{VT} \right)$.

Note 3: If the lower limit requirement of overturning moments is used in HR_DAD_RC by clicking the option of the requirement in the section 'Load Modeling' of the page 'Modeling', the calculated indexes are adjusted accordingly. |
| Plot response database | = Place the button of "Plot response database" at the bottom left corner of the page after a plot option (Bij_PM or Bij_VT) is chosen

→ The response database for the demand-to-capacity index of the selected member will be shown in the left side. The index is plotted as a function of wind speed and direction. |
| Plot peak response | = Place the button of "Plot response database" at the bottom left corner of the page after a member is selected on the member list

→ The peak response database of demand-to-capacity indexes for the member will be shown along MRIs. |

	Note: Peak demand-to-capacity indexes for any MRI can be obtained from this plot. The indexes shown in the plot correspond to the option in which the lower limit requirement for overturning moments is not selected.
Inter-story drift	
Once the button is clicked, the inter-story drift module is activated.	
Column set	= Select a column line set of interest from the list of column line sets, and click the button "Check" Note: the list of column line sets is obtained from "column lines of interest" in the page 'Wind Effects'.
Story	= Select a story of interest from the list of all stories of the structure, and click the button "Check" Note: the list of stories in the structure is obtained from "No. of stories" in the page 'Modeling'.
For MRI =	= Select a MRI of interest from the list of MRIs, and click the button "Calculate" → Inter-story drift ratio (x and y directions) of the selected story in the selected column line set for the selected MRI will be shown in the blanks. Note: The list of MRIs for inter-story drift is obtained from "MRIs for inter-story drift" in the page 'Wind Effects'.
Plot response database	= Place the button "Plot response database" at the bottom left corner of the page after a plot option (x axis or y axis) is chosen → The response database for inter-story drift ratio in the selected direction will be shown in the left side. The inter-story drift ratio is plotted as a function of wind speed and direction.
Plot peak response	= Place the button "Plot response database" at the bottom left corner of the page → The peak response database of inter-story drift ratios in both x and y directions will be shown along MRIs. Note: Peak inter-story drift ratio for any MRI can be obtained from this plot.
Acceleration on top floor	
Once the button is clicked, the top floor acceleration module is activated.	
Corner point	= Select a corner point of interest from the list of corner points, and click the button of "Check"

	Note: The list of corner points at top floor is obtained from "points on top floor of interest" in the page 'Wind Effects'.
For MRI =	= Select a MRI of interest from the list of MRIs, and click the button "Calculate" → Acceleration resultant at the selected point for the selected MRI will be shown in the blanks. Note: The list of MRIs for acceleration is obtained from "MRIs for acceleration" in the page 'Wind Effects'.
Plot response database	= Place the button "Plot response database" at the bottom left corner of the page after a plot option (x dir., y dir., or resultant) is chosen → The response database for top floor acceleration in the selected option will be shown in the left side. The top floor acceleration is plotted as a function of wind speed and direction.
Plot peak response	= Place the button "Plot response database" at the bottom left corner of the page → The peak response database of top floor acceleration in x dir., y dir., and resultant will be shown along MRIs. Note: Peak acceleration at top floor for any MRI can be obtained from this plot.

Load / Save / Exit button:

Load
A file containing input information for running HR_DAD_RC can be loaded by clicking the "Load" button located at the leftmost bottom.
Load a file containing input information for running HR_DAD_RC
→ All variables required for HR_DAD_RC can be loaded once they were previously saved.
The file is a MATLAB file generated by flnSAVE.

Save
The "Save" button specifies a file name and its location where the all input information loaded through pages one to seven will be saved.
→ All variables and filenames required for HR_DAD_RC can be saved once they were input in the user graphic interface mode.
The file is a MATLAB file that contains all the necessary information input at the time the file is saved.

Exit
The program can be stopped by clicking the "Exit" button located at the rightmost bottom.

References

ACI (2008). *Building code requirements for structural concrete (ACI 318-08) and commentary*, American Concrete Institute, Farmington Hills, MI.

ASCE (2010). *Minimum design loads for buildings and other structures (draft)*, American Society of Civil Engineers, Reston, VA.

Grigoriu, M. (2009). *Algorithms for generating large sets of synthetic directional wind speed data for hurricane, thunderstorm, and synoptic winds*. NIST Technical Note 1626, National Institute of Standards and Technology, Gaithersburg, MD.

Melbourne, W. H. (1980). "Comparison of measurements on the CAARC standard tall building model in simulated model wind flows." *Journal of Wind Engineering and Industrial Aerodynamics*, 6(1-2), 73-88.

Simiu, E., Gabbai, R. D., and Fritz, W. P. (2008). "Wind-induced tall building response: a time-domain approach." *Wind and Structures*, 11(6), 427-440.